Alternative Fusion Fuels and Systems

Alternative Fusion Fuels and Systems

S. V. Ryzhkov
A. Yu. Chirkov

CISP

CRC Press
Taylor & Francis Group
Boca Raton London New York

CRC Press is an imprint of the
Taylor & Francis Group, an **informa** business

Translated from Russian by V.E. Riecansky

CRC Press
Taylor & Francis Group
6000 Broken Sound Parkway NW, Suite 300
Boca Raton, FL 33487-2742

© 2019 by CISP
CRC Press is an imprint of Taylor & Francis Group, an Informa business

No claim to original U.S. Government works

First issued in paperback 2020

ISBN 13: 978-0-367-57077-4 (pbk)
ISBN 13: 978-0-367-02471-0 (hbk)

Visit the Taylor & Francis Web site at
http://www.taylorandfrancis.com

and the CRC Press Web site at
http://www.crcpress.com

Contents

List of abbreviations

CT   – Compact Toroid
CTC   – Compact Toroid Challenge
CTF   – Controlled Thermonuclear Fusion
DEC   – Direct Energy Conversion
DEMO – DEMO Power Plant
ECR   – Electron Cyclotron Resonance
FR   – Fusion Reactor
FRC   – Field-Reversed Conguration
FRE   – Fusion Rocket Engine
GDT   – Gas-Dynamic Trap
ICF   – Inertial Confinement Fusion
ICRH   – Ion Cyclotron Resonance Heating
IE   – inertial-electrostatic
ITER   – International Thermonuclear Experimental Reactor
ITF   – Inertial Thermonuclear Fusion
LDMIF – Laser-Driven Magneto-Inertial Fusion
LIF   – Laser Inertial Fusion
MAGO – Magnetic Compression
MCF   – Magnetically Confined Fusion
MeV   – Megaelectronvolt [1 eV = 11 604 K]
MFE   – Magnetic Fusion Energy
MGs   – Megagauss [1 T = 10^4 Gs]
MICF   – Magneto-Inertial Confinement Fusion
MIF   – Magneto-Inertial Fusion
MIIF   – Magnetically Insulated Impact Fusion
MTF   – Magnetized Target Fusion
NIF   – National Ignition Facility, USA
NPP   – Nuclear Power Plant
NS   – Neutron Source
OT   – Open Trap
PF   – Plasma Focus
PJMIF – Plasma Jet Driven Magneto-Inertial Fusion
QSNS   – Quasi-Stationary Neutron Source
RMF   – Rotating Magnetic Field
ST   – Spherical Tokamak

Notations

D   – deuterium

T   – tritium

^{3}He   – helium-3

^{4}He   – helium-4, or an α-particle

p – the proton

n – the neutron

^{11}B   – boron-11

e – electron

i – ion

pl   – the plasma

β – the ratio of the gas kinetic pressure of the plasma to magnetic pressure

β_B – Barnes beta

$\varepsilon_0 = 8.854 \cdot 10^{-12}$ F·m^{-1} – the electrical constant of the vacuum

λ_B – de Broglie wavelength

λ_D – Debye length

Λ_e – the electron Coulomb logarithm

Λ_i – the ion Coulomb logarithm

μ – the normalized mass of the ion, $\bar{\mu}$ is the average ion mass

$\mu_0 = 4\pi(10^{-7})$ [H ·m^{-1}] – the magnetic constant

ρ_{i0} – the ion synchrotron radius in the external field

σ – the parameter of the radial structure in FRC profiles

η_{syn} – conversion efficiency of synchrotron radiation

η_{therm} – thermal energy conversion efficiency

τ_E – the energy confinement time

τ_p – the confinement time of particles

τ_s – the slow-down time of fast ions

A – the area of the first wall

$B_c = B_e$ – the external magnetic field

e – the electron charge

$h = 6.626 \cdot 10^{-34}$ J·s – the Planck constant

m_e – the mass of an electron

m_i – the mass of the i-type ion

n_τ — coefficient of wave impedance/slow-down time
n_e — the electron density
$\langle n_e \rangle$ — the average electron density
n_i — the density of the i-type ion
$\langle n_i \rangle$ — the average ion density
n_0 — the total ion density
n_{prof} — density profile
P_{brems} — bremsstrahlung power
P_{elect} — fusion power transmitted to electrons
P_{ie} — the power transmitted in electron–ion collisions
P_{inj} — injection power
P_{syn} — synchrotron radiation
P_{th} — thermal power
P_{tot} — the total power of charged particle synthesis
P_q — the energy loss of charged particles
P_{wall} — neutron load on the wall
Q — the power gain factor in the plasma
r_c — the radius of the conducting wall
$r_s = a$ — the radius of the separatrix
r_w — the radius of the first wall
$L_s = b$ — the length of the separatrix
$\langle \sigma v \rangle_i$ — the average rate of the i-th reaction
T_e — the electron temperature
T_i — the ion temperature
Z_{eff} — the effective charge
Z_i — the charge of the i-type ion

Foreword

The present work is stimulated by the need to present the results of studies obtained by the authors in the field of alternative directions of thermonuclear fusion. It is addressed to specialists and researchers interested in this field of knowledge, and also the material of the book can be useful to graduate students who have chosen the topics related to high-temperature plasma as subjects for further scientific work. The authors hope that this publication will allow readers to expand their understanding of the range of issues related to the promising areas of the use of fusion energy.

Sections 1.4, 3.1, 3.4, 5.1, 5.2, 5.3 and Chapter 6 are written by S.V. Ryzhkov, sections 1.2, 1.3, 1.5, 2.1, 3.2, 3.3, 3.5 and Chapter 4 by A.Yu. Chirkov, introduction, foreword and sections 1.1, 2.2, 5.4 and 5.5 – jointly by the authors.

The authors express their sincere gratitude to Professor V.I. Khvesyuk, who is the permanent scientific supervisor of both authors and the inspiration for the work on low-radioactive fusion in Russia, as well as A.V. Anikeev, P.A. Bagryansky, I.Yu. Kostyukov, V.V. Kuzenov and I.V. Romadanov for consultations and discussions, as well as to his numerous colleagues for the full support and assistance in the preparation of this book.

Introduction

The solution of the problem of controlled thermonuclear fusion (CTF) is extremely important for the future energy sector, as already today there is an acute need for new sources of global energy. For the reaction of deuterium (D) with tritium (T), the conditions for implementing thermonuclear burn with a positive energy yield are the lightest among all known fusion reactions. Among stationary systems with magnetic plasma confinement, the tokamak occupies a leading position; followed by a stellarator. Other magnetic traps are commonly referred to as alternative directions.

The creation of a tokamak reactor with D–T fuel is now the main direction of the CTF program both in Russia and abroad [1]. At present, the international project of the experimental reactor-tokamak ITER (International Thermonuclear Experimental Reactor) [2] has entered the construction phase. Its main goal is physical demonstration of conditions corresponding to thermonuclear combustion with a power amplification factor in plasma $Q = 10$, which is a necessary condition for the efficiency of a thermonuclear reactor. At the final stage of ITER, work is supposed to be done on a D–T mixture. After ITER, it is planned to create a demonstration reactor (DEMO) [3] also based on a tokamak with D–T fuel. The task of DEMO is the development of technologies and engineering solutions.

It should be noted that impressive expectations are also associated with the achievements of the new large installation of laser inertial synthesis of NIF (National Ignition Facility, USA), but in this book the problems of inertial synthesis are not considered, but systems with magnetic confinement are analyzed.

If one adheres to the tokamak line with D–T fuel, then the beginning of the era of practical use of CTF is expected after 2050, when it is planned to create demonstration reactors [1]. Such a long period is required to a great extent to solve two of the most important tasks. The first is the construction of a large experimental ITER installation to demonstrate the regime with $Q = 10$. The second is the development of a blanket design for the reproducing of tritium and its testing in a DEMO reactor.

Very relevant is the search and justification for the possibility of faster introduction of thermonuclear devices into practical energy. From this point of view, thermonuclear systems that have significant potential advantages in comparison with the D–T tokamak reactor concept are promising.

One of such promising areas is the development of sources of thermonuclear neutrons for the disposal of long-lived radioactive waste and a hybrid fusion–fission reactor [4, 5]. In this case, the plasma gain $Q = 0.1$–1 is required. Therefore, the prototype of such a device can serve already existing installations, not only tokamaks. The creation of a larger experimental setup with $Q = 10$ is not required in this case.

An important task is to develop and justify the concept of a reactor using a mixture of deuterium and light helium isotope ^{3}He (helium-3) as fuel [6]. On Earth, helium-3 is practically absent (as well as tritium), but it is possible to extract it from the lunar regolith. In connection with the plans to create bases on the Moon and the prospects for industrial development of its subsoil, the solution of the problem of using helium-3 in power engineering can greatly enhance the economic feasibility of lunar projects. Therefore, it is necessary to have reasonable knowledge about the possibilities of creating a reactor on D–^{3}He fuel.

The most important advantage of the D–^{3}He reactor in comparison with the D–T reactor is the significantly reduced level of neutron radioactivity. The reactor on the D–^{3}He fuel is commonly called low-neutron, or low-radioactive.

It should also be noted that the obvious advantage of the reactor using only deuterium as primary fuel is the availability of fuel. The high efficiency of such a reactor can be achieved in the case of the so-called catalyzed D–D cycle, in which the products of the D–D reaction – tritium and helium-3 – participate in secondary reactions with deuterium. According to the level of neutron radioactivity, D–D fuel is comparable with D–T fuel, and plasma parameters and conditions for achieving high efficiency are close to the case of D–^{3}He fuel. Most of the neutrons produced in the D–D plasma have an energy inadequate for the driver of a hybrid reactor or nuclear waste utilization.

Yet another promising direction is neutron-free thermonuclear fusion. The most probable fuel for a neutron-free reactor is a mixture of protons (p) and isotope nuclei of boron-11 (^{11}B). The reserves of hydrogen and boron-11 on Earth are sufficient for the appropriate energy. The creation of a neutron-free reactor looks very attractive, but the low reaction rate of p–^{11}B, at first glance, does not leave a chance for a positive energy yield in such a reactor. Some hopes for improving the energy balance are associated with the possibility of maintaining a highly disequilibrium state, but this approach requires adequate analysis.

To date, there are conceptual designs for reactors on alternative fuels D–^{3}He, D–D and p–^{11}B, but as a rule, such projects consider the technical side, and many plasma parameters are based on considerations of a presumed nature. At least indisputable for a thermonuclear reactor with an alternative fuel is the need to confine plasma with $\beta \sim 1$ (β is the ratio of the plasma pressure to the magnetic pressure). Since in the classical tokamak $\beta \sim 0.1$, alternative fuels should consider alternative magnetic configurations in which $\beta \sim 1$ can be reached.

It is important to note that an increase in β with a constant value of the induction of the magnetic field makes it possible to increase the plasma density and the density of energy release. Consequently, systems with high β are in themselves of interest and deserve further development.

Many projects of neutron sources are oriented to tokamaks of the scale of today's experimental machines. Using a simple magnetic configuration with $\beta \sim 1$, for example an open trap, can significantly reduce the cost of the system and remove a number of engineering problems.

The presented research directions are aimed at the analysis of the following promising thermonuclear systems with high β: energy reactors on D–^{3}He fuel with $Q = 10$–20, neutron sources with $Q = 0.1$–0.5 and systems of magneto-inertial fusion.

To achieve maximum reliability of the results, the range of alternative configurations of the magnetic configurations (with respect to tokamak and stellarator) considered in this book is limited only by experimentally realized systems.

The creation of calculation and theoretical methods for analyzing the regimes of alternative magnetic thermonuclear systems is, in our opinion, necessary, since the available experimental data do not allow us to predict in full the parameters of thermonuclear plasma in such systems. For tokamaks, modelling is not as critical as there are empirical calculation techniques based on a huge amount of accumulated experimental data. It is also necessary to note the most important reasons why the experimental knowledge base on tokamaks can not be directly used for predicting thermonuclear regimes in prospective systems. First, these are significantly higher values of β in comparison with the limiting value $\beta \approx 0.1$ for classical tokamaks. Secondly, the high temperatures for D–^{3}Hefuel and the reaction p–^{11}B (~100 keV), much higher than the plasma temperature in the D–T reactor (~10 keV). Therefore, in order to achieve the goals set in the work, it is necessary to develop fundamentally new methods for analyzing processes in a plasma with allowance for high temperatures, high β, and extremely high plasma energy densities and magnetic fields.

Table 1 shows the reactions of nuclear fusion, which are of greatest interest from the point of view of energy production, having a high energy release and high rates.

The conditions for ignition of the D–T reaction are the easiest, which determines it as the unconditional leader of the fuel cycle of the first thermonuclear reactors. A serious drawback of the D–T reaction is high-energy neutrons, which account for 80% of the

Table 1. Basic and neutron-free (non-radioactive) thermonuclear reactions

Reaction (energy of products, MeV)	Energy release, keV	Direct radioactivity	Induced radioactivity
D + T → n (14.07) + ^{4}He (3.52)	17589	n	n, T
D + D → n (2.45) + ^{3}He (0.82)	3269	n	
D + D → p (3.02) + T (1.01)	4033	T	n
D + ^{3}He → p (14.68) + ^{4}He (3.67)	18353	–	n, T
D + ^{6}Li → 2 ^{4}He + 22.37	22371	–	n, T
D + ^{6}Li → p + T + ^{4}He + 2.257	2257	T	n
D + ^{6}Li → p (4.397) + ^{7}Li (0.628)	5025	–	
D + ^{6}Li → n (2.958) + ^{7}Be (0.423)	3381	n, ^{7}Be	
D + ^{6}Li → n + ^{4}He + ^{3}He + 1.794	1794	n	
D + ^{7}Li → n + ^{4}He + ^{4}He + 15,121	15121	n	
D + ^{7}Be → p + ^{4}He + ^{4}He + 16.766	16766	^{7}Be	
p + ^{6}Li → ^{4}He (1.7) + ^{3}He (2.3)	4018	–	n, T, ^{7}Be, 11 C
p + ^{11}B → 3 ^{4}He + 8.68	8681	–	n, ^{14}C
^{3}He+^{3}He → 2p (8.573) + ^{4}He (4.287)	12860	–	–

energy released. Today, there are no structural materials capable of retaining mechanical properties under neutron flux conditions on the first wall of the D–T reactor for more than 5 years. Since tritium is a rapidly disintegrating isotope (half-life of 12 years), a tritium-reproducing blanket is required to maintain the fuel balance of the D–T cycle. Development of blanket technologies in DEMO-reactor development programs takes more than 15 years. Another important factor, whose influence has increased especially now, is control over the non-proliferation of nuclear technologies. This can present certain obstacles to the deployment of D–T energy, since high-energy D–T neutrons are suitable for the production of nuclear materials. It is possible that the D–T reaction will be used in a controlled neutron source of a hybrid reactor in which the main energy is released upon fission of heavy isotopes in the blanket. Such reactors can be created practically at today's level of thermonuclear systems. Since the level of radiation hazard of a «clean» thermonuclear D–T reactor is comparable with the level of hybrid circuits (synthesis + division), then at the level of technical problems on the way to the implementation of the first hybrid systems can be more competitive.

In the D–^{3}He reaction, no neutrons are produced, which makes it potentially attractive from the point of view of a low-radioactive thermonuclear reactor. But it is impossible to organize a completely neutron production cycle on its basis, since in a plasma containing deuterium, two branches of the D–D reaction run in parallel, in which neutrons and tritium nuclei are produced. The latter, interacting with deuterium nuclei, yield D–T neutrons. Thus, the D–^{3}He cycle includes the first four reactions from Table 1, among which D–^{3}He reaction – the main for the power. The energy output in neutrons is from 3 to 10%, depending on the share of burn of tritium and other parameters. At the level of neutron fluxes from the plasma of the D–^{3}He reactor, the service life of the first wall is about 30 years, that is, practically equal to the lifetime of the reactor. Today, the absolute leaders of thermonuclear systems are tokamaks, both in terms of achievements and in the costs of research. Achieving high energy production efficiency in the D–^{3}He reactor based on the classical tokamak is limited by low values of the parameter β. For the D–^{3}He reactor, it is desirable β; $\beta \gtrsim 0.5$, while, for example, in tokamaks $\beta \sim 0.1$. High β is necessary to reduce cyclotron losses (due to a decrease in the magnetic field in the plasma) with technically achievable reflection coefficients of synchrotron radiation by the wall.

It should be noted that the direction of hybrid fusion–fission reactors is very promising now. Thermonuclear systems can be used to destroy long-lived radioactive waste from fission reactors.

In alternative reactors with low-radioactive (neutron-free) fuel, the production of tritium and, consequently, the subsequent production of weapons-grade nuclear materials.

The main problems of thermonuclear installations are now heating and long-term stable plasma confinement, ensuring plasma cleanliness, and reducing turbulent losses. For the under construction and designed installations, powerful additional heating systems are required, among which the main ones are high-frequency, beam and ohmic heating. To heat the plasma, the technologies of radio emission sources are developed – electron cyclotron resonance heating (ECR-heating) and ion cyclotron resonance heating (ICR – Ion Cyclotron Resonance), as well as heating by a beam of neutral atoms (NBI – Neutral Beam Injection). Plasma confinement in modern experiments only on single installations reaches 1 s. The criterion for igniting a D–T reaction at a temperature $T \sim 10^8$ K is the Lawson condition $n\tau > 10^{20}$ m^{-3}·s, where n is the density of the plasma, and τ is the confinement time. The ratio β of the plasma pressure p to the magnetic pressure $B^2/2\mu_0$ in the tokamak is small (~ 0.1). The plasma density and, consequently, the volume density of energy release in a tokamak are limited. Following the dependence $\tau \sim a^2$, where a is the radius of the plasma, we can say that with an increase in the size of the tokamak it is possible to increase $n\tau$, but one can not achieve a greater specific energy release. In alternative magnetic configurations with a high plasma pressure ($\beta \sim 1$), the required value of the Lawson parameter can be achieved at relatively small sizes.

Magnetic configurations with high β and high energy density, such as an open magnetic trap, field reversed configuration (FRC) spheromak, and others, represent alternative systems (with respect to tokamak) that can be used for applications of nuclear fusion.

Systems of magnetic confinement of high-density plasma [7] are of interest for improved (low-radioactive) cycles (a small fraction of the thermonuclear power attributable to neutrons) and for the concept of magneto–inertial fusion (MIF).

The advantages of systems with low radioactive fuel and high β are the low neutron load, high energy density and the absence of a blanket for the production of tritium, which simplifies the task of creating an alternative thermonuclear reactor.

Since the primary work of O.A. Lavrent'ev on thermonuclear energy and A.D. Sakharov and I.E. Tamm on the magnetic thermal insulation of high-temperature plasma (the latter proposed tokamak system, called by I.N. Golovin from the word combination 'TOroidal'naya KAmera i MAgnitnye Katushki'), Soviet and Russian scientists invariably remained at the forefront of the development of magnetic confinement systems for plasma. It should be noted that the greatest contribution to the formation and development of compact systems was made by Russian researchers and engineers.

Many systems of magnetic confinement of high-temperature plasma were proposed by Soviet scientists. An open trap was proposed by G.I. Budker (independently of R. Post and H. York), followed by an axisymmetric ambipolar trap (G.I. Dimov), a gas-dynamic trap (D.D. Ryutov). Much has been done by American, European and Japanese scientists.

It was on the basis of a number of the above facilities that the diffusion pinch, traps with a reversed field, etc., were later created, as well as the recently received interest in the system of magnetic-inertial fusion (MIF, magneto-inertial fusion or MTF – magnetized target fusion). These include, for example, the system with the liner compression of magnetized plasma, MAGO (abbreviation for MAGnitnoe Obzhatie) in Russia and MTF in the US, which combine the approaches of magnetic and inertial plasma confinement.

The main experimental setups, their location and main goals are given in Table 2 in alphabetical order. In addition, there are other facilities, such as Rotamak (Flinders University, Australia), SSX (Swartmore, Pennsylvania), TS-3, TS-4 (Tokyo University), in which FRC is formed by the merging of spheromaks. The ranges of experimentally obtained plasma parameters in compact tori and magnetic traps are presented in Table 3.

And although at the present time most of the efforts of the magnetic thermonuclear community are devoted only to one main application – the creation of a thermonuclear power station, but given the well-developed nuclear power in the 21st century, it is unlikely that they will be able to be countered by a nuclear power plant. In this connection, it is necessary to find applications for thermonuclear neutrons and, first of all, they can serve as a driver for subcritical nuclear power systems – a fusion hybrid reactor. Such a system can also be used for transmuting long-lived radioactive waste and generating nuclear fuel for auxiliary nuclear systems. A small thermonuclear system based on FRC is well studied for

Table 2. Alternative magnetic traps – experiment

FRC	Open traps	Spheromaks
CBFR – Univ. of California, Irvine p–^{11}B	GDL, SHIP – Budker INP SB RAS, pressure increase, β	BCTX – CI, Berkeley, heating
FIREX – Cornell University, Ithaca Munsat, Univ. of Colorado, Bouldder	CLM – Columbia University, instability	BSX – injection of compact tori, Caltech
FIX – Osaka University; NUCTE-3 Nihon University	GAMMA 10 – PRC, University of Tsukuba	HIT-CT – Himeji, Japan
FRX-L – LANL, MIF/MTF compression, high density	Gasdynamic trap – LLNL, tokamak feed, large beta	CTIX – University of California, Davis, acceleration
KT, TOP, TOP-Liner – TRINITI, compression, transfer, compression	GOL-3 Multipurpose trap – Budker INP SB RAS,	HIT-SI – University of Washington, a new way of forming
FRC in FIAN. P.N. Lebedev RAS – formation, large capacity	FLM – Uppsala University	SPHEX – UMIST, poloidal and toroidal fields
MRX, PFRC – Princeton, form, stability, reconnection	HANBIT – Korean Institute of Basic Sciences, Daejeon	SSPX – LLNL, Livermore, retention
TCSU, STX, TRAP, PHD, IPA – rotating field, growth of T, flow	MultiCusp Trap – RRC Kurchatov Institute	SSX – Swarthmore, multi-probe reconnection
C-2 – Tri Alpha Energy, Calif., Merging 2 CTs	MAP-II – University of Tokyo	TS-3,4 – Tokyo, FRC other tor. configuration

these purposes. The characteristics of the source of thermonuclear neutrons based on it and the potential for reprocessing radioactive nuclear waste and the production of nuclear fuel in the blanket are presented in [8].

Table 3. Experimental parameters of the plasma

Parameter	FRC	Open traps	Spheromaks
Radius of the separatrix r_s, m	0.02–0.40	0.02–0.40	0.01–0.30
Length l_s, m	0.2–1	0.25–12	0.2–0.7
Electron density n_e, m^{-3}	0.005–5·10^{22}	10^{16}–7.5·10^{22}	0.001–1·10^{20}
Ion temperature T_i, keV	0.05–3	0.03–10	0.05–0.5
Electron temperature T_e, keV	0.03–0.5	0.004–4	0.02–0.5
External magnetic field B_e, T	0.05–2	0.005–15	0.03–3
Average beta $\langle \beta \rangle$, %	75–95	2–70	5–20
Energy confinement time τ_E, ms	0.05–0.5	0.01–1	0.02–2

Alternative systems of magnetic confinement of plasma and low-radioactive (or neutronless) fuels [9, 10] can not only have the above applications, but also be applied in other areas. At present, experimental installations of this direction can already find application as prototypes of fast proton and neutron sources, for studying the interaction of plasma with a wall, for testing the main most energetic elements of structures. In the near future, applications of high-temperature FRC plasma using deuterium and light helium (D–^{3}He), proton and boron (p–^{11}B) and other cycles of thermonuclear reactions can be used. Systems for supplying fuel in the form of a field reversed magnetic configuration or spheromak are considered,

Table 4. Main characteristics of existing MITS units and their comparison with MTS and ITS

Systems / Parameters	Magnetic confinement	MAGO	FRC	Inertial confinement
Magnetic field, T	0.01–20	10	4	—
Density, m^{-3}	10^{20}	10^{24}	1023	10^{27}
Temperature, eV	20000	300	$T_e + T_i \sim 500$	Cryogenic
The plasma lifetime, s	1000	10^{-5}	2· 10–5	~ ns
Synthesis energy, MJ	~1 GW	Liner ~ 20 MJ		~5
Liner speed, m/s	–	10000		100000
Pulse power	150 MW, 25 MA (ITER)	10 MJ, 50 MA		1.8 MJ Laser (NIF)
Estimated price	~ $10·10^9	~ $50 million		~$1 · 10^9
Beta plasma	<1	~1	~0.92	—

detection of explosives and chemical wastes, cancer therapy and production of medical isotopes: $^{18}O + p \rightarrow n + {}^{18}F$; $^{94}Mo + p \rightarrow n + {}^{94}Tc$; $^{14}N + p \rightarrow {}^4He + {}^{11}C$; $^{16}O + p \rightarrow {}^4He + {}^{13}N$; $^{13}C + p \rightarrow n + {}^{13}N$; $^{15}N + p \rightarrow n + {}^{15}O$. They are also discussed for their utilization of nuclear materials and radioactive waste.

The formation of FRC and spheromak is also extremely important for understanding the dynamics of plasma in the solar corona and astrophysical processes. Thus, the proposed compact tori together with low-neutron fuel are an unlimited source of clean energy and various applications in the future.

Other applied areas of compact systems are also very attractive. They can be used as a source of protons and neutrons, in material technology, for studying the interaction of a plasma-wall, for charging the tokamaks and for testing the most loaded structural elements.

Understanding the processes occurring in reversed magnetic traps, combined (magneto–inertial) configurations, hybrid (fusion–fission) systems, and the role of low-radioactive fuels are critically important for the future energy sector.

The parameters of the systems of the three concepts of controlled fusion for comparison are presented in Table 4.

Various projects of private companies of prototypes of fusion reactors are given in Table 5.

Table 5. Main parameters, reactions and types of plasma confinement in the planned reactors

Company	Diameter of plasma, m	Hold	Reaction	Power, MW
Convergent Scientific	~10	IE	p–^{11}B	225
Energy/Matter Conversion Corporation (EMC2)	~ 10	IE	p–^{11}B	100
General Fusion	3	MI	D–T	100
Helion Energy (MSNW)	16	MI	D–D	50
Lockheed Martin Skunk Works	~10	M	D–T	100
Lawrenceville Plasma Physics (LPP)	~ 0.15	PF	p–^{11}B	5
Magneto-Inertial Fusion Technologies Inc. (MIFTI)		MI		
Sorlox	~1	MI	D–D	~1
Tri Alpha Energy	18	MI	p–^{11}B	100

Table 6. Properties of compact torus configurations

Plasmoid	Axial symmetry	Poloidal magnetic field	Toroidal magnetic field
FRC	Yes	Yes	No
Spheromak	Yes	Yes	Yes

The first two settings are stationary, the rest – to impulse systems. Types of plasma confinement: IE – inertial electrostatic, MI – magnetically inertial, M – magnetic confinement, PF – plasma focus. Let us note that these projects are aimed not only at the development of power plants, but also in medical equipment.

The presence of poloidal and toroidal components of the magnetic field in the configurations of a compact torus (FRC) and a combined compact torus (spheromak) are presented in Table 6.

Low-radioactive fuel cycle based on the D–³He reaction

1.1. Advantages and possibilities of using helium-3 as a thermonuclear fuel

Studies on controlled thermonuclear fusion (CTF), conducted in many countries of the world over the past five decades, have now come very close to implementing a reactor design capable of demonstrating the possibility of industrial application of the energy of fusion of light nuclei. In 2005, the countries participating in the ITER (International Thermonuclear Experimental Reactor) project decided to build a reactor. In this reactor, the thermonuclear plasma will be held by a magnetic field in a tokamak type system, and the fuel cycle is based on the reaction of deuterium with tritium

$$D + T \rightarrow n(14.1 \text{ MeV}) + {}^{4}He(3.5 \text{ MeV}).$$

The values in the parentheses are the energies of the products–particles in megaelectronvolts (for energy $1 \text{ eV} = 1.6 \cdot 10^{-19}$ J). This reaction among all of possible thermonuclear reactions proceeds with the greatest rate and gives a positive energy yield at the lowest values of the plasma temperature. The plasma temperature in the ITER will be at the level of 15 keV (for a temperature of $1 \text{ eV} \approx 11\ 600$ K).

A serious problem in the way of creating a D–T reactor for industrial energy is due to the fact that in conditions of powerful neutron fluxes, modern materials can last only a few years. The service life of the corresponding structural elements is only 4–6 years instead of the required 30–50 years. At the same time from the

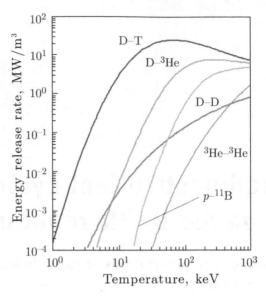

Fig. 1.1. The rates of energy release in some thermonuclear reactions.

technical point of view, the fusion reactor is much more complicated than the fission reactor. This leads to a sharp rise in the cost of energy produced. Therefore, fuel cycles alternative to deuterium-tritium can be more suitable for future industrial thermonuclear energy.

Figure 1.1 shows the rate of energy release in some thermonuclear reactions in 1 m^3 of plasma at a characteristic density of reacting nuclei of 10^{20} particles/m^3.

A possible alternative is the reaction of deuterium with a light isotope of helium

$$D + {}^3He \rightarrow p(14.68 \text{ MeV}) + {}^4He(3.67 \text{ MeV}).$$

All elements of the corresponding reaction are non-radioactive, and among the reaction products there are no neutrons. This leads to a significant reduction in the induced radioactivity of structural materials and, accordingly, to an increase in the lifetime of the reactor equipment, especially the first wall. In addition, the generated charged particles transmit their energy directly to the plasma before they leave it.

Intensive studies of the issue of energy production in the thermonuclear D–^{3}He cycle were started in the late 1980s in the USSR, the United States and Japan. Then the question arose about obtaining the isotope ³He, which is practically absent on the Earth. Researchers from the University of Wisconsin (USA) put forward the concept of extraction of helium-3 on the Moon, the reserves of which can be enough for mankind for about 1000 years. To meet the needs of the world's energy sector, it is enough to deliver about 100 tons of helium-3 per year from the Moon.

Recently, in the United States, it was decided to finance the lunar base program and extraction of helium-3 for energy needs. Plans for the extraction of lunar helium-3 were also announced by Russia and China. This means that in the near future these countries need to create a D–³He reactor project with technical elaboration at the ITER level.

The attractiveness of the D–³He reaction is primarily due to the fact that among its products there are no neutrons and radioactive isotopes, but it must be taken into account that parallel to the D–³He reaction in the plasma, there are reactions between the deuterium nuclei

$$D + D \rightarrow n(2.45 \text{ MeV}) + {}^3\text{He}(0.817 \text{ MeV}),$$

$$D + D \rightarrow p(3.02 \text{ MeV}) + T(1.01 \text{ MeV}),$$

in which neutrons and radioactive tritium are produced. Since the rates of D–D reactions are small compared to the rate of the D–³He reaction, a thermonuclear fuel cycle based on the D–³He mixture is considered to be slightly radioactive. According to estimates, in the D–³He cycle the neutrons account for about 5% of the energy released.

For effective burning of D–³He fuel, a temperature of the order of 70 keV is required, which significantly exceeds the burning temperature of the D–T reaction. Figure 1.1 shows that at temperatures of 100 keV or more, the energy release rates in D–T and D–³He approach each other. However, it should be borne in mind that as the temperature rises, the intensity of the electromagnetic radiation of the plasma (in the radiofrequency and hard X-ray ranges) and the energy losses associated with radiation increase. At 15 keV, the radiation losses from a pure D–T plasma account for several percent of the energy released from the fusion. For D–³He plasmas at 70 keV, radiation can take away more than half of the energy. Hard X-ray radiation is the inevitable loss channel. The radiofrequency losses can be significantly reduced by the diamagnetic weakening of the magnetic field in a high-pressure plasma. Due to the tokamak characteristics, the plasma pressure in it is small, and therefore it is hardly possible to achieve high energy efficiency of the D–^{3}He fusion. Thus, the tokamak is a system only for the D–T reactor. For alternative thermonuclear cycles, it is necessary to consider from all existing systems those that allow one to hold the plasma at pressures close to the physical limit.

In the discussions on thermonuclear fuel cycles, the neutron-free reaction of a proton and boron-11 is often considered

$$p + {}^{11}\text{B} \rightarrow 3\,{}^4\text{He} + 8.681 \text{ MeV.}$$

For its implementation, it is necessary to raise the fuel temperature above 200 keV. In addition, estimates have shown that the radiation losses for a stationary boron–proton plasma under ideal conditions are somewhat higher than the released energy. This reaction, of course, deserves attention, since for it on the Earth there are the necessary reserves of fuel, but based on today's thermonuclear technologies, there are no prospects for its application in the energy sector.

In the light of the use of lunar helium-3, some researchers mention a neutronless reaction

$$^3\text{He} + {}^3\text{He} \rightarrow 2p + {}^4\text{He} + 12.86 \text{ MeV.}$$

The rate of energy release in it is extremely low (see Fig. 1.1), which also excludes it from the field of energy applications of thermonuclear fusion. Thus, the most realistic prospects for the development of thermonuclear power with reactors of low neutron activity are associated with the D–^{3}He cycle, provided that sources of the ^{3}He isotope are available in near space or the generation of this isotope on the Earth is realized. Note that the group of the Bauman Moscow State Technical University, headed by Professor V.I. Khvesyuk, developed and invetigated D–^{3}He cycles with helium-3 production. The possibility of such cycles is achieved at the cost of increasing the energy yield in neutrons, therefore, in order to minimize neutron activity, delivery of helium-3 from the Moon is necessary.

As noted above, a high-pressure thermonuclear reactor is required for the D–^{3}He mixture. The systems of magnetic confinement of plasma are divided into two categories. In the first category, the magnetic field lines are closed in the confinement region.

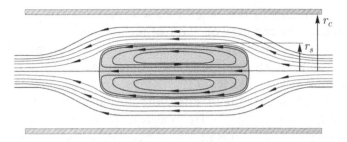

Fig. 1.2. The FRC configuration – the plasma inside the separatrix and the magnetic flux.

Representatives of this group are tokamaks, stellarators, spherical tokamaks. Another class includes systems in which magnetic lines go beyond the confinement region (to infinity). An example of such devices is field reversed (magnetic) configurations, spheromaks, ambipolar reactors and gas-dynamic traps.

The most suitable are compact systems: a spherical tokamak and a reversed magnetic configuration (FRC, field reversed configuration). The difficulties of using a spherical tokamak for D–³He fusion are associated with very high parameters and technological problems for this system, due to the presence of a rod in the centre of the installation. FRC is more attractive from a technical point of view (Fig. 1.2).

FRC refers to alternative thermonuclear circuits. This direction was distinguished from straight-line, i.e., non-toroidal pinch systems, however, by the method of confining the plasma, the compact torus belongs to closed magnetic traps.

In this compact configuration, the plasma is confined in equilibrium in a closed field and is separated from the conductive wall by an area of the open magnetic flux. A separatrix separating closed and open magnetic force lines is the natural boundary of a hot plasma. The internal closed field is maintained by currents in the plasma, while the outer region is vacuum or is filled with low-pressure plasma. The configuration is formed by reversal (change of direction) of the magnetic field.

Unlike tokamak, FRC has a simple cylindrical geometry that allows direct conversion of fusion energy into electricity with high efficiency. The structure of the magnetic field FRC is formed not only by currents in the linear system of external coils, but also by the diamagnetic azimuthal current in the plasma. Most importantly, the plasma pressure in the FRC is extremely high. This makes it possible to obtain in the reactor based on FRC a high density of energy release in comparison with other magnetic systems.

Table 7 shows the possible parameters of a reactor based on FRC operating on a D–³He fuel, and also for comparison the parameters of a tokamak-reactor ITER.

From a technical point of view, the system looks like an almost ideal candidate for the D–³He reactor. To date, not all physical processes in the FRC plasma have been investigated in the same detail as in the tokamaks. One of the most important problems that require further investigation is associated with the anomalous mechanisms of transport of particles and energy of the plasma across the magnetic field and the potential possibilities of its suppression.

Thermonuclear systems based on FRC, taking into account today's technologies, have prospects for using in several directions besides thermonuclear power plants: electric propulsion, isotope production for medicine, production of hydrogen fuel and a number of other technologies in various fields of science and technology.

It should be noted that even with the price of the helium-3 isotope at $1 million per ton, its production on the Moon and delivery to the Earth is energy equivalent to oil at a cost of about $10 per barrel.

Table 7. Parameters of FRC reactor with D–³He-fuel and ITER reactor

Reactor	ITER Tokamak	FRC
Fuel	D–T	D–³He
Thermonuclear power, MW	500	1900
The ratio of thermonuclear power to the input heating power	10	20
Form and dimensions of plasma	torus, small radius 2.1 m, large radius 6.2 m	cylinder, radius 1.2 m, length 20 m
Plasma temperature, keV	13	72
Magnetic field of coils, T	5.7	6.4
Maximal ratio of plasma pressure to external magnetic field pressure	0.1	1
The energy lifetime of the plasma, s	6	1.5
Energy output in neutrons,%	80	5

The total energy consumption for the ³He delivery is estimated at $2.4 \cdot 10^6$ MJ/kg. If it is considered that when helium is burned in a thermonuclear reaction $6 \cdot 10^8$ MJ/kg is released, then the energy gain is 250 times. This figure should be compared with the fact that the burning of uranium in nuclear reactors will increase energy by 20 times, and when burning coal – 16 times.

Note that the price of fuel (the cost of ³He, obtained from lunar soil, is assumed to be equal to $1,000,000/kg) is about 10% of the cost of electricity, which includes waste disposal costs. Further, with the deepening of research and the operation of industrial thermonuclear power plants, the technology will become cheaper, as it is now, for example, with computer systems.

1.2. Alternative deuterium cycles

Below we consider the characteristics of thermonuclear fuel and the corresponding fuel cycles, without being tied to any particular system. In the initial analysis, the plasma parameters will be assumed constant throughout the volume.

From the point of view of the energy efficiency of thermonuclear cycles, the most important characteristics are the ratio of power, released in neutrons, to thermonuclear power (neutron yield)

$$\xi_n = P_n / P_{\text{fus}} \qquad (1.1)$$

and the ratio of the bremsstrahlung power to the thermonuclear power

$$\xi_b = P_b / P_{\text{fus}}. \qquad (1.2)$$

Neutrons and bremsstrahlung are not absorbed in the plasma, and therefore the energy carried away by them is inevitably lost from the plasma.

Figure 1.3 shows the location of various cycles in the diagram $\xi_b - \xi_n$ [11]. In the designations of cycles in the diagram, the first symbols indicate the primary reaction, the symbols following them indicate which primary reaction products enter secondary reactions. For example, D–D means an uncatalyzed cycle in which the possibility of participation of products of the main D–D reaction in the secondary reactions D–T and D–³He is not taken into account. A fully catalyzed cycle in which tritium and helium-3, which are produced in the main D–D reaction, completely react with deuterium,

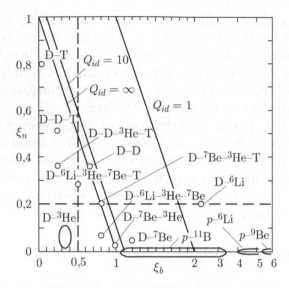

Fig. 1.3. The location of various thermonuclear fuel cycles in the $\xi_b - \xi_n$ diagram.

is designated D–D–³He–T. In the partially catalyzed D–D–³He cycle, only secondary helium-3 burning is considered. In cycles based on the reactions of D–³He, D–⁶Li and D–⁷Be, the contribution of D–D reactions is taken into account, since D is the main component of the fuel along with ³He, ⁶Li, and ⁷Be.

To illustrate the energy efficiency, the ξ_b–ξ_n diagram uses an ideal gain

$$Q_{id} = \frac{P_{fus}}{P_n + P_b - P_{fus}} = \frac{1}{\xi_n + \xi_b - 1}, \qquad (1.3)$$

taking into account only losses with neutrons and bremsstrahlung.

In cycles lying on the diagram to the left of the $Q_{id} \to \infty$ line, it is possible to ignite at a finite value of the confinement time. The requirement of reduced neutron activity in the diagram corresponds to the region $\xi_n < 0.2$. The acceptable level of losses for bremsstrahlung is $\xi_b < 0.5$. Since the ideal case of pure plasma is considered, then taking into account impurities in the reactor, the value of ξ_b can be higher.

Proton cycles are practically neutron-free, which makes them attractive from the point of view of radioactivity of the reactor. But due to high losses to bremsstrahlung, the prospects for their use require special analysis.

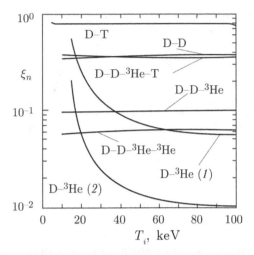

Fig. 1.4. Neutron yield in deuterium thermonuclear cycles: D–T ($n_D = n_T$); D–³He ($n_D = n_{3He}$, *1* – tritium burns completely, *2* – tritium does not burn); pure and catalyzed D–D.

As can be seen from the diagram, quite a lot of variants of alternative deuterium cycles based on the D–⁶Li, D–⁷Be reactions have a low neutron yield $\xi_n < 0.2$, but only D–³He cycles are located in the $\xi_b < 0.5$ region. Thus, among the alternative thermonuclear cycles, the cycles with the highest energy efficiency correspond to those in which the primary fuel components are deuterium and helium-3.

Let us consider in detail the question of radioactivity. In deuterium cycles, D–D neutrons with an energy of 2.45 MeV are produced, and in the case of secondary tritium burn D–T neutrons with an energy of 14.1 MeV are also produced. In addition, in cycles with lithium-6 neutrons with energies of 2.96 MeV and about 0.66 MeV are produced in two of the five branches of the D–⁶Li reaction.

The values of the neutron yield $\xi_n = P_n/P_{fus}$ as a function of temperature $T_i = T_e$ for cycles based on the reactions D–D, D–T and D–³He are shown in Fig. 1.4. The cycle designated D–D–³He–³He assumes that the resulting tritium is completely removed, and then its decomposition takes place with conversion to helium-3. The amount of helium-3 in this cycle corresponds to its operating time in the D–D reaction and the disintegration of the tritium produced.

As can be seen in Fig. 1.4, in alternative deuterium cycles, even with complete tritium burnout, the neutron yield is several times lower than in the D–T cycle. The average neutron energy is also several times lower than the energy of D–T neutrons.

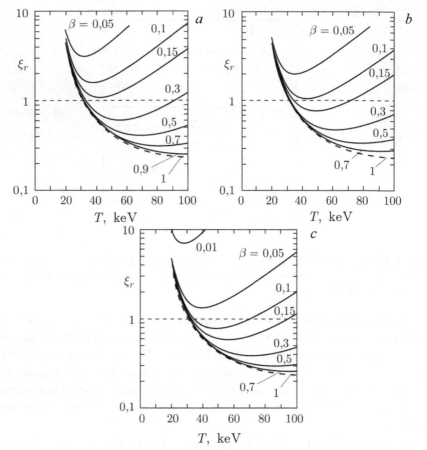

Fig. 1.5. The relative magnitude of the radiation losses for the D–^{3}He cycle at $n_{^3\text{He}} = n_\text{D}$, $T_e = T_i = T$, $R_w = 0.8$ and the different values of β, the vacuum magnetic field B_0 and the plasma radius a: $a - B_0 = 5$ T, $a = 1$ m; $b - B_0 = 5$ T, $a = 4$ m; $c - B_0 = 20$ T, $a = 4$ m.

When analyzing catalyzed cycles, we consider two limiting cases: complete burning of tritium in a catalyzed cycle and so-called partially catalyzed cycles in which tritium is completely removed from the plasma. The problem of tritium burnout and the associated D–T neutrons exists for all deuterium alternate cycles, since the rate of the D–T reaction is high and a significant amount of tritium formed will react with deuterium and give 14 MeV neutrons.

Let us analyze the radiation losses. The smallest amount of radiation losses corresponds to bremsstrahlung (P_b). Since the operating temperatures are relatively high for the D–^{3}He cycle and other alternative cycles, a significant contribution to radiation losses is made by synchrotron radiation.

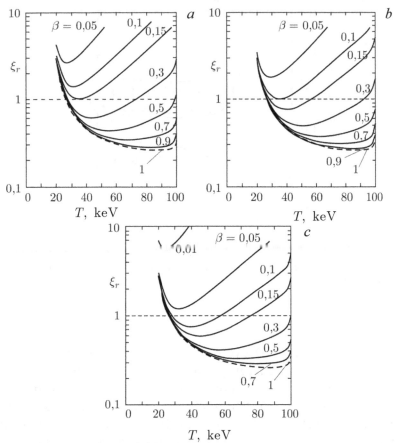

Fig. 1.6. The relative magnitude of the radiation losses for a fully catalyzed D–D cycle at $T_e = T_i = T$, $R_w = 0.8$ and different values of β, the vacuum magnetic field B_0 and the plasma radius a: a – $B_0 = 5$ T, $a = 1$ m ; b – $B_0 = 5$ T, $a = 4$ m; c – $B_0 = 20$ T, $a = 4$ m.

Synchrotron losses (P_s) depend on the electron temperature, the induction of the vacuum magnetic field B_0, the parameter β, the radius of the plasma column a, and other factors. The relative magnitude of the radiation losses

$$\xi_r = (P_b + P_s)/P_{fus} \tag{1.4}$$

is shown in Fig. 1.5 and 1.6 for the D–³He cycle and the fully catalyzed D–D cycle. According to Fig. 1.5, for the D–³He cycle with reasonable magnetic fields and plasma sizes, taking into account $\xi_n \approx$ 0.05–0.1, the thermonuclear power can exceed losses at $\beta > 0.15$–0.3.

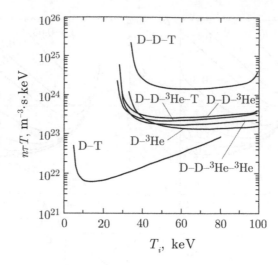

Fig. 1.7. The ignition criterion for deuterium cycles: D–T ($n_D = n_T$), D–^{3}He ($n_D = n_{3He}$) and catalyzed D–D-cycles.

For the catalyzed D–D cycle, $\xi_n \approx 0.35$, and, according to Fig. 1.6, the excess of the energy yield over the losses is possible for $\beta > 0.3$–0.5. Taking into account other losses, the requirements for the value of β for D–^{3}He and catalyzed D–D cycles are still increasing. Therefore, when considering reactors with magnetic confinement for these cycles, it is necessary to focus primarily on systems in which confinement with $\beta \approx 0.5$ or more is possible in principle.

The ignition in ideal conditions ($Q_{id} \to \infty$) can be characterized by the modified Lawson criterion $n\tau T$, where n is the total concentration of plasma particles (ions and electrons), τ is the energy confinement time, and T is the plasma temperature ($T = T_i = T_e$ is assumed). The criterion $n\tau T$ makes it possible to estimate the minimum requirements for a given fuel type for the confinement system. The values of the criterion $n\tau T$ as a function of temperature for thermonuclear cycles based on deuterium, helium-3 and tritium are shown in Fig. 1.7.

The ignition criterion values are close to D–^{3}He and catalyzed D–D cycles. For cycles based on the reactions of D–^{6}Li and D–^{7}Be they are about an order of magnitude higher. Another order of magnitude higher values of $n\tau T$ are necessary for proton cycles, even with $Q_{id} \leq 1$.

1.3. A low-radioactive D–³He cycle with ³He production

Let's consider the main properties of the D–³He cycle, characterizing the requirements for plasma confinement and the radioactivity of energy production, depending on the content of ³He in a thermonuclear plasma [12]. The analysis must be carried out taking into account all the reactions taking place simultaneously in the deuterium-containing plasma (see Table 1 of the introduction):

$$D + {}^3He \rightarrow p(14.68 \text{ MeV}) + {}^4He(3.67 \text{ MeV}), \qquad (1.5)$$

$$D + D \rightarrow n(2.45 \text{ MeV}) + {}^3He(0.817 \text{ MeV}), \qquad (1.6)$$

$$D + D \rightarrow p(3.02 \text{ MeV}) + T(1.01 \text{ MeV}), \qquad (1.7)$$

$$D + T \rightarrow n(14.1 \text{ MeV}) + {}^4He(3.5 \text{ MeV}). \qquad (1.8)$$

Because of the high rate of reaction (1.8), a significant fraction of the tritium nuclei produced in the reaction (1.7) have time to react with deuterium before leaving the trap.

The parameters of the cycles based on deuterium reactions (1.5)–(1.8) are largely determined by the ratios between the concentrations of the plasma components

$$x_j = n_j / n_D, \qquad (1.9)$$

where j = ³He, T.

Figure 1.8 shows the minimum values of the ignition criterion $n\tau T$ as a function of the relative content of helium-3 $x_{{}^3He} = n_{{}^3He}/n_D$. The temperature T_{min} corresponding to the minimum of $n\tau T$ for a given $x_{{}^3He}$ was determined during the calculation. The values of $n\tau T$ in Fig. 1.8 correspond to the temperature T_{min}, which is shown in Fig. 1.9.

The relative content of tritium x_T, which determines the intensity of the flux of 14 MeV neutrons, depends on which part of the tritium has time to react with deuterium during the confinement time. Burnout of tritium in a plasma is characterized by the ratio γ_T of the number of tritium nuclei burning per unit time to the number of tritium nuclei produced per unit time.

In a low-radioactive thermonuclear reactor, the neutron yield must be ensured at the level $\xi_n \sim 0.1$, which requires minimizing the neutron flux with an energy of 14.1 MeV. To do this, it is desirable

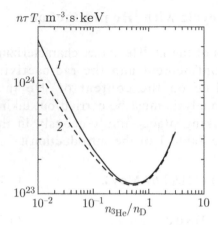

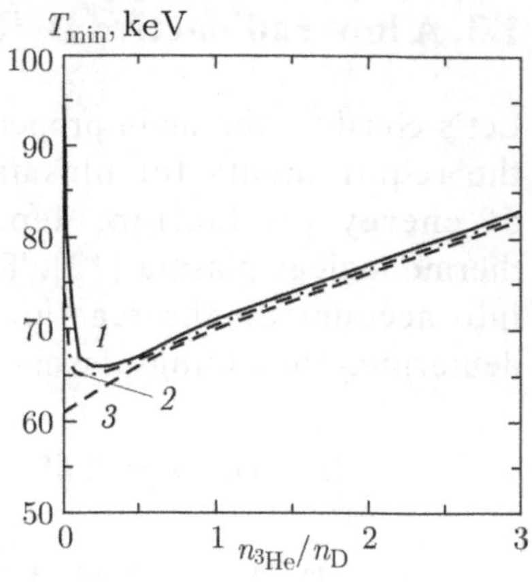

Fig. 1.9. Temperature corresponding to the minimum value of the ignition criterion $n\tau T$ as a function of the relative amount of ^{3}He in the D–^{3}He cycle: tritium produced in reaction (1.7) does not burn (*1*); 50% of tritium burns (*2*); tritium burns completely (*3*).

Fig. 1.8. Ignition criterion $n\tau T$ as a function of the relative content of ^{3}He in the D–^{3}He cycle at $T = T_{min}$: tritium produced in reaction (1.7) does not burn (*1*); tritium burns completely (*2*).

to force tritium removal from the plasma, which can be done, for example, by the so-called selective pumping [13]. At the same time, other reaction products are also removed: p, ^{3}He, ^{4}He.

The fraction of the tritium burnout γ_T and the neutron yield ξ_n are shown in Figs. 1.10 and 1.11. To ensure $\xi_n = 0.1$–0.15, a relative amount of helium-3 $x_{3_{He}} > 0.2$ is necessary.

An acceptably low neutron yield $\xi_n \sim 0.1$ can be ensured not only in the equicomponent D–^{3}He cycle, but also in a cycle with a relatively small content of light helium $x_{3_{He}} \sim 0.2$, which can be achieved by producing ^{3}He directly in the operation of the D–^{3}He reactor. Such a reactor uses only deuterium as a raw material (primary fuel).

The operating time of helium-3 implies that the light helium obtained in a certain way is accumulated, and then sent to the plasma, where the reaction (1.5) is used to generate energy.

The following variants of operation of helium-3 are possible:

1) the reaction (1.6);

2) the collection of tritium, obtained in the reaction (1.7) and not having time to react with deuterium, its further exposure in the storage to conversion into helium-3 according to the reaction $T \rightarrow {}^3He + e^- + 0.018$ MeV;

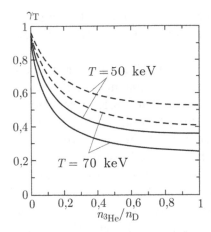

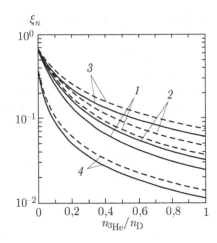

Fig. 1.10. The fraction of tritium burnout in the D–³He cycle at $\tau_T = \tau_{ign}$ (solid) and $\tau_T = 2\tau_{ign}$ (dashed curves); τ_T is the confinement time of tritium, τ_{ign} is the time corresponding to the ignition criterion $n\tau T$.

Fig. 1.11. Neutron yield in the D–³He cycle at $T = 70$ keV (solid) and $T = 50$ keV (dashed curves): $1 - \tau_T = \tau_{ign}$, $2 - \tau_T = 2\tau_{ign}$, 3 – complete burning of tritium, 4 – complete removal of tritium.

3) decay of tritium generated in the reactor blanket as a result of neutron reactions with lithium.

In addition, neutron multiplication is possible in the blanket with beryllium:

$$^9\text{Be} + n \rightarrow 2\,^4\text{He} + 2n - 1.67 \text{ MeV}, \tag{1.10}$$

which makes it possible to increase the production of tritium.

The efficiency of tritium production in the blanket is characterized by the coefficients K_{D-D} and K_{D-T}, which are the ratios of the number of nuclei T produced to the number of neutrons produced in the plasma.

Thus, the production of ³He is possible due to the reactions (1.6), (1.7) in the plasma and reactions in the blanket.

The common for all D–³He cycles with ³He production is that as raw material (primary fuel) they use only comparatively available deuterium, and synthesized during operation of the reactor ³He and T are used as secondary fuel. Below are the variants of cycles considered in [11, 12].

1. The cycle with the production of ³He and T in the plasma. In this case, a part of the light helium and tritium formed in the reactions (1.6) and (1.7) burns in the reactions (1.5) and (1.8). The remaining part, leaving the plasma, is extracted from the gas mixture

pumped out by the vacuum pumps, stored, after which ^{3}He, obtained in reaction (1.6) and as a result of transmutation of tritium, is sent to the plasma.

2. A cycle with the production of ^{3}He and T in plasma and the selective removal of fusion products from the plasma. In this case, it is assumed that all the products of thermonuclear reactions, including ^{3}He and T, slow down in the plasma, giving up their energy, and then, after reaching an energy of the order of several hundred kiloelectronvolts, are forcibly removed from the plasma. The rejected ^{3}He and T are stored. Accumulated helium-3, including as a result of the decay of tritium, is used as a component of the D–^{3}He fuel. This cycle has important advantages. A significant part of tritium does not have time to interact with deuterium in the reaction (1.8). This will make it possible to obtain a greater amount of ^{3}He than in the first variant. It is important that the neutron flux to the first wall is significantly reduced in comparison with the first case. And the decrease is due to the most dangerous fast neutrons with an energy of about 14 MeV.

3. A cycle with the development of tritium in a blanket. Neutrons produced in reactions (1.6) and (1.8) can cause the formation of tritium when ^{6}Li, ^{7}Li and ^{9}Be (for reproduction) are present in the blanket. In this variant, as in the previous two, also ^{3}He and T, produced in plasma, are also used.

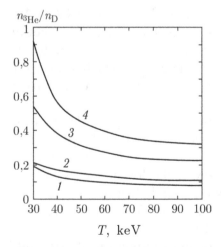

Fig. 1.12. Amount of ^{3}He in the cycle: *1* – ^{3}He, obtained in the D–D reaction; *2* – ^{3}He, obtained in the D–D reaction and in the decay of secondary tritium (without a blanket); *3* – the same with a blanket, $K_{D-D} = K_{D-T} = 1$; *4* – the same with the blanket, $K_{D-D} = K_{D-T} = 2$. The confinement time of tritium $\tau_T = \tau_{ign}$.

4. Variants that combine cycle 3 with cycles 1 or 2 are also possible.

Figure 1.12 presents achievable values of $x_{^3He}$ for different cycles. It can be seen that achievable $x_{^3He}$ values in the operating temperature range $T = 50\text{-}70$ keV range from 0.1 to 0.3, which is quite sufficient to provide highly effective low-radioactive energy production in the D–³He cycle, with self-sufficiency in the isotope ³He.

The following is common for all possible D–³He cycles with an excess of deuterium. Reactions of deuterium with deuterium serve mainly for the production of ³He and T (including due to the produced neutrons), which ultimately provides the cycle with the ³He isotope. The main production of energy is the reaction of deuterium with light helium (1.5).

The results of calculations of the thermonuclear regimes of D–³He cycles with respect to reactors with magnetic confinement are presented in Table 8 for relatively small and large β.

The analysis is based on an integrated balance of capacity. It is assumed that the plasma temperature is constant over the cross section of the plasma column, and the pressure has a parabolic profile

$$p = p_0[1-(r/a)^2], \tag{1.11}$$

Table 8. Parameters of D–³He plasma in a reactor with magnetic confinement at $P_{fus0} = 6$ MW/m³, $a = 2$ m, $\tau_a = 8$ s, $Q = 20$

	$\langle\beta\rangle = 0.1,$ $T = 50$ keV, $R_w = 0.9$		$\langle\beta\rangle = 0.5, T = 70$ keV, $R_w = 0.7$			
Option	1	2	3	4	5	6
$K_{D-D} = K_{D-T}$	—	1	1	0	0	1
$x_{^3He} = n_{^3He}/n_D$	0.5	0.305	0.245	0.125	0.158	0.245
$x_T = n_T/n_D$	0.0036	0.0038	0.0052	0.0057	$<10^{-3}$	$<10^{-3}$
γ_T	0.64	0.68	0.64	0.70	—	—
n_{D0}, 10^{20} m⁻³	2.55	3.11	2.51	3.27	3.26	2.70
n_{tot0}, 10^{20} m⁻³	9.72	9.87	7.68	8.38	8.06	7.38
B_0, T	10.7	10.7	5.4	5.4	5.4	5.3
τ_E, s	7.6	7.9	5.3	7.5	4.9	4.2
ξ_n	0.10	0.15	0.15	0.27	0.061	0.042
ξ_b	0.36	0.32	0.24	0.25	0.23	0.22
ξ_s	0.21	0.21	0.061	0.065	0.064	0.059
J_n, MW/m²	0.20	0.30	0.30	0.53	0.12	0.08

where r is the distance from the axis of the plasma column, and a is the radius of the column.

The specified parameters are the thermonuclear power on the axis P_{fus0}, the plasma radius a, the confinement time of the thermonuclear products τ_a, the power amplification factor in the plasma Q, the average value $\langle \beta \rangle$, the plasma temperature $T = T_i = T_e$, the reflection coefficient of the cyclotron radiation by the wall R_w, tritium production rates in the blanket K_{D-D} and K_{D-T}.

Table 8 shows the calculated $x_{3_{He}}$, x_T the tritium burnout γ_T, the deuterium density on the axis n_{D0} axis, the total density of all plasma particles on the axis n_{tot0}, the vacuum value of the magnetic field induction B_0, the required energy confinement time τ_E, $\xi_n = P_n/P_{fus}$, $\xi_b = P_b/P_{fus}$, $\xi_s = P_s/P_{fus}$ and the energy flux of neutrons from the plasma J_n.

Option 1 with $x_{3_H} = 0.5$ corresponds to the smallest $n\tau T$ (Fig. 1.9). The relative content of helium $x_{3_{He}} = 0.5$ is provided by using external sources of ^{3}He. Calculations of variants 2–6 are performed in the approximation of the steady-state helium-3 production, when the burning and production rates are equal. The variants 2 and 3 are the cycles with the production of helium-3 in plasma and tritium in plasma and blanket. The differences in $x_{3_{He}}$ in these variants are related to the temperature dependence of the reaction rates (1.5)–(1.8). In the cycles 3–5, the first and second methods of obtaining ^{3}He are realized, in cycle 6 – a combination of the second and third methods.

In the variants 5 and 6, a pumping is assumed, which ensures the complete absence of tritium with an energy of <350 keV. It is assumed that the density of thermalized tritium is zero. The density of high-energy tritium is calculated from the high-energy distribution function [14]. The energy losses from the plasma, associated with the pumping out of thermonuclear products, slowed to 350 keV, account for about 10% of P_{fus}.

Note that the D–^{3}He cycle is sensitive to contamination by impurities entering the plasma when the first wall coating is sprayed. For example, at $x_{3_{He}} = 0.15–1$ and $T = 50–70$ keV, the maximum permissble content of Be^{+4} is 2–3%, B^{+5} is 1–2%.

The cycles with selective pumping of thermonuclear products are most attractive among the variants considered above. Pumping allows to minimize the neutron yield ξ_n to a level 2–4 times lower than for the cycles without pumping. The flux of neutrons to the first wall is minimal, and the radiation losses ξ_b and ξ_s are minimal. It is also important that in such cycles the minimum energy confinement time τ_E is required.

The cycles based on the reactions of D–³He and D–D, apparently, are of greatest interest for CTF among the alternative cycles. For $\beta \sim 1$ systems with such cycles can compete with a traditional D–T reactor. Prospects of energy production in alternative cycles depend on the capabilities of containment systems with $\beta \sim 1$: the field reversed configuration, the spherical tokamak, the open traps, and others.

The conditions for achieving high energy efficiency (ignition criterion, operating temperature, neutron yield, etc.) are close to D–³He and catalyzed D–D cycles. The analysis presented above showed that it is possible to implement an intermediate version – D–³He cycle with ³He production, in which the main part of the energy produced is accounted for by the D–³He reaction, and only available deuterium is used as the primary fuel. In this case, there is no need to deliver helium-3 from the Moon.

The need for manipulation with tritium and the creation of neutron shielding exists for the D–³He cycle both with the ³He lunar source and with the ³He production. In the latter case, the use of a blanket for the production of tritium in reactions involving neutron capture for the purpose of subsequent production of ³He in the decay of tritium may be justified, since the amount of ³He in the cycle increases, which leads to a decrease in the neutron yield.

1.4. Possible types of reactors based on deuterium–helium-3

The D–T reaction predominates in studies of thermonuclear fusion, since it has the fastest rate. However, D–³He and p–¹¹B reduce neutron production and are environmentally friendly thermonuclear products. Private venture capital firms such as Tri Alpha Energy [15], Helion Energy [16], General Fusion Inc. [17] and EMC2 Energy Matter Conversion Corporation [18] are interested in them. The fuel p–¹¹B has no limitations, but at a temperature corresponding to the maximum energy release rate (~300 keV), the energy losses to bremsstrahlung are so great that it is impossible to obtain a positive energy yield.

The D–³He fuel cycle represents a compromise variant – much cleaner than D–T (almost like the neutron-free p–¹¹B reaction), and it is possible to maintain plasma combustion with slightly greater difficulties than D–T. But on Earth there is a relatively small amount of helium-3 (^{3}He) isotope, mainly from nuclear reactors in heavy water, decomposition of tritium in nuclear warheads and natural gas. Large reserves of ^{3}He on the Moon and in the Solar System, lunar helium-3 are a matter of growing international interest [19, 20]. Although at the moment there is not enough of ^{3}He as fuel for engineering tests.

Thermonuclear power engineering based on the D–^{3}He reaction will contribute to the non-proliferation of nuclear weapons [10, 21]. Preliminary analysis shows that the operational regimes of the D–^{3}He FRC thermoelectric power stations occur at very low levels of neutron production and, therefore, the production and use of nuclear fuel is virtually impossible. This situation is directly opposite to D–T thermonuclear reactors, in which every 3–5 years the internal parts (first wall and 30–50 cm deep) must be changed, which makes the power plant vulnerable to illegal module replacements, which can lead to substitution fuel that undergoes fission and nuclear fission.

Along with the main reactions listed above, there are promising thermonuclear reactions, sometimes called exotic ones, on the basis of which it is possible to organize neutron-free cycles of energy production.

The conditions for ignition of the D–T reaction are the easiest, which determines it as the unconditional leader of the fuel cycle of the first thermonuclear reactors. A serious drawback of the D–T reaction is high-energy neutrons, which account for 80% of the energy released. Today, there are no structural materials capable of retaining mechanical properties under neutron flux conditions on the first wall of the D–T reactor for more than 3–5 years. Since tritium is a rapidly disintegrating isotope (half-life of 12 years), a tritium-reproducing blanket is required to maintain the fuel balance of the D–T cycle. Development of blanket technologies in the DEMO reactor development programs takes more than 15 years [1]. Another important factor, whose influence has increased especially now, is the control over the non-proliferation of nuclear technologies. This can present certain obstacles to the deployment of D–T energy, since high-energy D–T neutrons are suitable for the production of nuclear materials. Perhaps the D–T reaction will be used in a controlled neutron source of a hybrid reactor [5, 22], in which the main energy is released upon fission of heavy isotopes in the blanket. Such reactors can be created practically at today's level of thermonuclear systems. Since the level of radiation danger of a 'clean' thermonuclear D–T reactor is comparable with the level of hybrid circuits (fusion + division), then at the level of technical problems on the way to the implementation of the first hybrid systems can be more competitive.

In the reaction (1.5) neutrons are not produced, which makes it potentially attractive from the point of view of a low-radioactive thermonuclear reactor. But it is impossible to organize a completely neutrone production cycle on its basis, since in the plasma containing deuterium the reactions (1.6) and (1.7) in which neutrons and tritium nuclei are generated are parallel. The latter, interacting with deuterium nuclei, yield D–T neutrons. Thus, the D–³He cycle includes the reactions (1.5)–(1.8), among which the reaction (1.5) is the main one with respect to the released power. The energy output in neutrons is 3–10%, depending on the fraction of combustion of tritium and other parameters. At the level of neutron fluxes from the plasma of the D–³He reactor, the service life of the first wall is about 30 years, that is, practically equal to the lifetime of the reactor.

In the Earth's interior and atmosphere, the necessary reserves of the isotope ³He are absent. Sufficient reserves for the corresponding energy are available in the Moon's soil (regolith). If the recently announced plans of a number of countries (China, the European Union, Russia, the United States, Japan) to create bases on the Moon and industrial development of its subsoil begin to be realized, then in the near future there will be an urgent need for an industrial D–³He reactor project. Today, the absolute leaders of thermonuclear systems are tokamaks both in terms of achievements and in the costs of research. Achieving high energy efficiency in the D–³He reactor based on the classical tokamak is limited by low values of the parameter β (β is the ratio of the plasma pressure to the magnetic pressure).

For the D–³He reactor, it is desirable to have $\beta \gtrsim 0.5$, while, for example, in tokamaks $\beta \sim 0.1$. High β is necessary to reduce cyclotron losses (due to a decrease in the magnetic field in the plasma) with technically achievable reflection coefficients of cyclotron radiation by the wall.

If we do not count on the delivery of helium-3 from the Moon, we can consider catalyzed D–D cycles as an alternative. They also include the reactions (4.9)–(4.12). In this case, the primary fuel is only readily available deuterium, there is no need for a fuel-reproducing blanket. The energy output in neutrons is from 30 to 35%, which is comparable to the D–T reactor. Neutron energies are insufficient for use in hybrid circuits.

Variants of the D–^{3}He cycle are possible with the production of helium-3 in the reactor [11, 12]. Such cycles combine both the production of helium-3 in the plasma, and the operating time in the decay of tritium, obtained in the blanket. It is possible to organize a cycle with the main reaction (4.10) with a yield level in neutrons of 10–15%. The minimum neutron yield can be obtained by removing tritium from the plasma using a selective pumping system. Such cycles in principle allow solving the problem of the first wall. A significant drawback is the need to manipulate significant amounts of tritium.

The reaction p + ^{11}B $\rightarrow$ 3^{4}He + 8.681 MeV makes it possible to organize a neutron-free energy production cycle provided with fuel resources. Unfortunately, due to the low speed of the p–^{11}B reaction, the prospects for its effective use can not be fully determined with today's level of knowledge. At least for the p–^{11}B reactor with magnetic confinement, a trap with $\beta \sim 1$ is needed. There are several more reactions (see Table 1 of the introduction), which in principle can be considered in comparison with the reactions (4.10)–(4.12) [11]. These are deuterium reactions with lithium-6 (neutron yield in D–^{6}Li cycles at the catalyzed D–D cycle level, ignition conditions are tougher), deuterium with beryllium-7 (^{7}Be is radioactive, rapidly decays), neutron-free proton reactions with ^{6}Li and ^{9}Be (the rates are much lower than the p–^{11}B reactions), the reaction ^{3}He + ^{3}He $\rightarrow$ p + p + ^{4}He + 12.86 MeV. The latter is relatively often discussed in connection with the problem of neutron-free fusion, but its rate is extremely low.

In addition, it is possible to combine the above reactions. When using the p–^{6}Li reaction, a D–^{3}He–^{6}Li cycle can be performed. The advantages and disadvantages (direct and induced radioactivity) of all reactions are shown in Table 9. The reaction of deuterium with helium-3 has the maximum specific power of energy release after the deuterium–tritium reaction. A positive output for igniting the reaction and maintaining the burn (power gain factor $\sim$10) in the D–^{3}He plasma can be achieved at $\sim$50 keV. Thus, for non-radioactive thermonuclear reactors D–^{3}He fuel looks most promising.

When analyzing the efficiency of D–^{3}He reactors, these models require mathematical models that combine the most detailed description of various processes in the plasma with respect to the magnetic configuration in question. Such integrated models of thermonuclear plasma were created at the Bauman Moscow State Technical University. Numerical codes were created on their basis, which allow calculating both elementary processes at the kinetic level or the level of individual particles, as well as the integral parameters of the plasma and the magnetic system of the reactor [23, 24]. For the operating temperatures of reactors on alternative fuels, calculations of the bremsstrahlung of relativistic electrons are performed and approximating formulas are obtained. To calculate the heating of plasma components by thermonuclear products and injected fast particles, a kinetic model was developed based on the Fokker–Planck equation, taking into account Coulomb and elastic nuclear collisions, as well as the participation of products in secondary reactions [25]. For the calculation of anomalous transport, models for the transport of individual particles under the action of specified perturbations [26], electromagnetic drift instabilities [27], as well as nonlinear saturation of drift turbulence structures have been developed.

Energy balance calculation methods were used to analyze D–^{3}He reactors with magnetic confinement [24, 28–39]. The main parameters of the reactors are given in Table 9. We considered D–^{3}He cycles with different values of the ratio of the concentrations of helium-3 and deuterium $x_{^3\mathrm{He}} = n_{^3\mathrm{He}}/n_\mathrm{D}$. For example, $x_{^3\mathrm{He}} \approx 0.3$ corresponds to the maximum helium-3 operating time in the reactor. In the case of $n_{^3\mathrm{He}} = /n_\mathrm{D}$ the typical value of the neutron energy flux to the first wall is $q_n \approx 0.15$ MW/M^2.

A high-β system using an alternative fuel (with respect to D–T) has better parameters, for example, a low neutron yield and a high energy density. Because of the small neutron load, waste in low-radioactive systems will have a low radioactivity and, consequently, a lower cost of installation. The optimal size of the neutron-free thermonuclear power station is only slightly more than the thickness of the blanket, protection and coil combined. Figure 1.13 shows geometries for conventional and low-radioactive fuels: a design comparison for D–T (left) and D–^{3}He (right) reactions. The cylindrical shape of the chamber and the single-connected nature of the plasma body greatly simplify the magnetic system, the chamber design and many other engineering problems that are critical for modern toroidal systems.

The cylindrical geometry helps in transporting the FRC along the axis and moving it to a special combustion chamber remote from the most vulnerable radiation zone of formation. This also simplifies the design of the blanket. Comparison of the cross section of the D–T reactor with the D–^{3}He system leads to the conclusion that, other conditions being equal, the plasma radius for the D–^{3}He reactor

Table 9. Parameters of D–^{3}He reactors

Type of reactor	Tokamak [28]	Spherical tokamak [29]	Stellara-tor [30]	FRC [31]	CT
Radius of plasma a, m	2	3	2	1.6	1.25
Length of plasma L, m	—	—	—	35	30.75
Aspect ratio	3	1.5	20	1	1
The elongation of the plasma E	2.5	3.8	1	11	12.3
The external magnetic field B_0, T	11.3	3.2	8.2	5	6.4
Current in plasma I, MA	38	200	—	—	299
The mean β	0.09	0.54	0.1	0.46	0.75
Fuel composition $x_{3\text{He}} = n_{3\text{He}} / n_D$	0.2	0.36	1	1	1
Aaximum plasma temperature T, keV	50	60	70	60	72
The coefficient of reflection of synchrotron radiation by the wall Γ_s	0.92	0.65	0.95	0.5	0.99
Confinement time τ, s	14	16	30	4	1.46
Thermonuclear power P_{fus}, MW	2500	1500	1500	1000	1937
Relative power of bremsstrahlung P_{br}/P_{fus}	0.40	0.60	0.15	0.53	0.38
Relative power of synchrotron radiation P_s/P_{fus}	0.33	0.023	0.25	0.06	0.005
Relative power in neutrons P_n/P_{fus}	0.12	0.15	0.02	0.07	0.025
Gain $Q = P_{fus}/P_{ext}$	20	20	∞	20	40

will be smaller than for the D–T. And this, in turn, ensures the non-proliferation of weapons of mass destruction, that is, the inability to produce radioactive isotopes in the case of a D–^{3}He mixture. The smaller size and absence of zones and elements characteristic

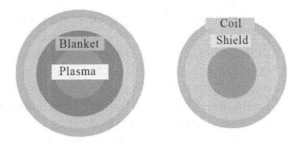

Fig. 1.13. Structure of the reactor zone (schematic section of the combustion chamber) for D–Ti and D–³He plasmas. The main elements that determine the size of the magnetic confinement system and the impossibility of generating radioactive elements and the non-proliferation of weapons of mass destruction are shown.

of the D–T reactor, make it impossible to produce nuclear fuel in the blanket. Comparison with the projected D–³He-Tokamak-based reactor shows that the D–³He FRC-based systems will have better characteristics.

The stellarator reactor has substantially larger dimensions and a higher thermal power. If the small cross-section has dimensions close to other reactors, then the aspect ratio and, consequently, the large radius of the torus is almost 7 times larger than in the tokamak reactor. A significant increase in the aspect ratio is caused by the need to reduce the high plasma losses predicted by the neoclassical transfer theory for stellarators. In terms of the cost of such a project, the stellarator reactor loses to other systems, since the cost of the system is proportional to its volume.

However, if we consider the prospects for industrial production of energy, then such systems may be more acceptable. Unlike other types of reactors, in this project the fusion reaction is maintained stationary by heating the plasma with reaction products, and therefore the reactor does not require powerful stationary plasma heating systems. Heating of the plasma is only necessary to bring the reactor into the combustion mode. The stationary combustion eliminates the problems associated with the heat pump surrounding the plasma reactor design elements, which reduces the time resource of the system and can trigger the emergence of emergency situations. The problem of reliability of the operation of all reactor systems as a powerful energy system is extremely important.

The positive features of the stellarator reactor should also include the stability of the combustion modes of the thermonuclear mixture, as was shown for the case of the D–T reaction in [34].

The stellarator makes it possible to choose plasma parameters such that the dependences of the transport coefficients responsible for diffusion and thermal losses have an inverse relationship to the collision frequency. The presence of such regimes is characteristic of stellarator systems. In this case, a random increase in the plasma temperature is extinguished by higher energy losses, and its decrease leads to a decrease in thermal losses and returns the temperature to its original position. One of the problems of powerful energy devices, such as thermonuclear reactors, is the problem of possible malfunctions, the elimination of which requires reactor shutdown and repair work inside the vacuum chamber or in adjacent systems. To carry out such work may require significant periods of time, which is associated with large economic losses. The modular design of the stellarator reactor makes it possible to simplify such a problem. Replacing a faulty module with a spare, serviceable module can significantly shorten the repair time. There are a number of proposals for modular designs of stellarator reactors. Presented in Table 9 version of the reactor also assumes its modular design. From the point of view of the immediate prospects for the development of a thermonuclear reactor, the stellarator obviously can not compete with other simpler devices because of its scale. At the same time, it remains one of the promising devices for creating industrial reactors.

The stellarator and compact torus have the highest plasma temperatures and the lowest value of the bremsstrahlung fraction. They also have the minimum power in neutrons in comparison with thermonuclear power and the maximum power amplification factor. In this case, a compact torus with a purely poloidal magnetic field (inverted magnetic configuration) has smaller dimensions (a compact system) and a higher value of β. If the average in the spherical tokamak is limited, then in compact tori it can practically reach a value of ≈ 1.

Table 10 presents four conceptual projects: Apollo [40] and ARIES-III [41] – Tokamaks on D–^{3}He fuel, Artemis [42] and FRC RV proposed by the author [43] – D–^{3}He-FRC-based reactors. From Table 10 it can be seen that the D–^{3}He FRC power plants have large powers of charged particles and β, smaller values of injection power and magnetic field. The advantages of the D–^{3}He tokamaks are low combustion temperatures and long confinement times. The proposed variant has a smaller neutron flux than the traditional closed systems of magnetic confinement of high-temperature plasma (tokamak and stellarator), and a lower plasma temperature than in the Artemis

Table 10. Comparison of the main parameters of D^{-3}He-reactors based on tokamak and FRC (conceptual designs)

System characteristics	Apollo [40]	ARIES-III [41]	Artemis [42]	FRC RV [43]
Electric power, MW	1000	1000	1000	1000
Fusion power, MW	2144	2682	1610	1962
Bremsstrahlung, MW	652	Radiation fraction 0.72	Radiation 357	776
Synchrotron radiation, MW	1027			8.7
Charged particles, MW	456		1181	1188
Neutrons, MW	147	110	77	51.7
Injection (current maintenance), MW	(138)	(172)	5	62.6
Efficiency of the system	0.43	Recycling 0.24	0.36–0.62	0.49
Neutron flux, MW/m^2	5.7	Average 0.08	0.27	0.15
Fuel ^{3}He/D	0.63	~1	0.5	1
Large radius (separatrix length), m	7.89	7.5	(17)	(30.75)
Small radius (radius of the separatrix), m	2.5	2.5	(1.12)	(1.23)
Ion temperature, keV	57	55	87.5	68.5
Electron temperature, keV	51	53	87.5	68.5
Electron density, m^{-3}	$1.9 \cdot 10^{20}$	$3.3 \cdot 10^{20}$	$6.6 \cdot 10^{20}$	$5.4 \cdot 10^{20}$
Density of ions, m^{-3}	$1.3 \cdot 10^{20}$	$2.1 \cdot 10^{20}$		$3.46 \cdot 10^{20}$
Plasma current, MA	53	30	160	298.8
Toroidal magnetic field on the axis (external), TI	10.9 (19.3)	7.6	(6.7)	(6.38)
The mean β,%	6.7	Toroidal 24	90	74.8
Energy confinement time, s	16	11.8. $\tau_p^{ash} / \tau_E = 2$	2.1. $\tau_p/\tau_E = 2$	1.44

project. Although plasma current is of great importance, but one must understand that this current flows axially along the device. For this reason, the current/elongation ration I/E is the best indicator for estimates when creating and maintaining current. For this option $I/E = 24$ MA. A similar value, for example, in a thermonuclear

D–^{3}He reactor based on a spherical tokamak $I \sim 73$ MA [44], in which case the thermonuclear power is 1400 MW. Technology issues are not included in this review, but we note that the use of liquid lithium (near-wall pumping) [45] as the material of the first wall of an industrial power plant leads to a minimum value of the recycling coefficient [46] and the wall structure behind a liquid seven times the neutron mean free path will withstand for 30 years at a load of 30 MW/m^2 and more [47].

The parameters of the D–^{3}He reactors obtained in the calculations, in our opinion, are adequate to the potential capabilities of the corresponding systems. The most preferred system is the reversed magnetic configuration (FRC), which has a relatively simple magnetic system from a technical point of view. The serious problem of FRC today is the lack of a sufficient level of understanding of the mechanisms of anomalous transport. The validated concepts of the D–^{3}He-reactor are possible for a spherical tokamak and a stellarator, since the processes in these systems obey the laws well studied in 'classical' tokamaks and stellarators.

The ratio of the concentrations of helium-3 and deuterium in fuel cycles affects both the released thermonuclear power and the energy yield in neutrons. Thus, in the equicomponent D–^{3}He reaction, the fraction of neutrons is ~5%, which is ~15% lower than in a mixture with a predominance of deuterium. Thus, the minimum energy yield in neutrons in D–^{3}He-cycles is reached only when using lunar helium-3. It is important that in this case the first wall of the reactor does not need to be replaced during the whole lifetime of the reactor, nor does it require a blanket for any reproduction of the fuel. The latter circumstance makes it possible to significantly shorten the time for the creation of a demonstration reactor in comparison with a reactor having a blanket.

The parameters of D–^{3}He reactors were compared on the basis of existing magnetic confinement systems: a tokamak, a spherical tokamak, a stellarator, a reversed magnetic configuration, or an elongated compact torus. The most attractive from a technical point of view is FRC (compact toroid (CT), which is due to the simplicity of the linear geometry of the external magnetic field. One of the advantages of the D–^{3}He reactor based on the reversed magnetic configuration is the high density of energy release, which makes it possible to create a relatively compact reactor of the same power as the deuterium–tritium tokamak. A high efficiency of the D–^{3}He reactor based on a spherical tokamak requires extremely high values

Table 11. Parameters of plasma for different thermonuclear fuels (projects of magnetic systems of confinement of hot plasma)

Project	ITER toka-mak	KSTM trap	EPSILON (TMT)	Levi-tating dipole	FRC	CBFR (CT)
Fuel	D–T	D–T	D–D	D–D	D–^{3}He	p–^{11}B
Thermo-nuclear power, MW	500	100–500		610	1900	8.68
Power gain	> 10	1.5–5.0	Stationary	1.5 (T/ Radia-tion)	40	2.65 (T/Radia-tion)
The size and the shape of the plasma, m	R/a = 6/2, elonga-tion 17	r = 1.5, l = 30	1.6 × 5.2, d = 0.3 in the sole-noid	β – 10.3 at max. pressure	r_s = 1.25, l_s = 25	ellipsoid r_0 = 0.4
Volume, m³	840	106 (212)	0.8–1.2	269 (PS)	120	
Plasma tempera-ture, keV	13	T_i = 15–60 T_e = 60–150	100– 500 eV	T_i = 41, T_e = 30	72	T_i = 235. T_e = 82
Magnetic field, T	5.7, po-loid. 1	3 in the center. cell	0.1 — solenoid 1 — CREL*	30 in the reel	6.4	1.53
The maxi-mum β	0.1	0.3	0.1	0.031	1	1
Confine-ment time, s	6	~ 1	1–10	5.1	1.5	Disintegration 36
Neu-trons,%	80	70	30–40	40	2.5	0

* CREL is a curvilinear equilibrium element.

of the magnetic field for this system. A significant advantage of the spherical tokamak is that the behaviour of the plasma in it obeys the same laws as in the classical tokamak for which the level of research is significantly ahead of all other containment systems. Regimes for efficient energy production are found, comparison of conceptual designs of D–³He reactors based on tokamak and compact torus is given.

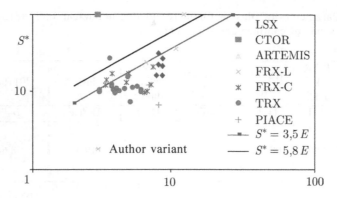

Fig. 1.14. Modes of operation of the experimental FRC and the conceptual design of the FRC reactor with D–^{3}He fuel. The parameter S^*, depending on the elongation of the configuration E.

Comparison of the compact thermonuclear D–^{3}He FRC reactor [48] with the ITER tokamak [49] kinetically stabilized by the open KSTM trap [50], a pseudosymmetric system (PS) containing toroidal mirror traps (TMT) without rotational transformation of EPSILON [51], a levitating dipole [52] and a compact torus [53] with different fuels is given in Table 11. There are also other versions of thermonuclear power plants, for example, based on a long antiprobe trap [54], Z-pinch [55, 56] and other schemes [57].

The dependence of the dimensionless parameter S^* on the elongation of FRC is shown in Fig. 1.14 (data for analysis are taken from [58]).

From the point of view of the energy efficiency of a thermonuclear reactor, the greatest interest is in the production of energy and, correspondingly, the plasma power amplification factor Q, which is equal to the ratio of the power released as a result of the fusion reactions to the additional power injected into the plasma to maintain the specified temperature of the fuel. To create radiation-safe thermonuclear installations with a self-sustaining combustion reaction, it is necessary to satisfy the condition $Q > 10$.

FRC, using a low-radioactive (with respect to D–T) fuel, has a low neutron yield and a high energy density. Improved confinement in the FRC opens the way to using this configuration as a fusion reactor, and this can improve the attractiveness of other FRC applications [38, 59, 60], such as the tokamak makeup or the use of FRC for the fusion of a magnetized target fusion (MTF). It is necessary to continue the research in close contact with the experiments, for example, the C-2 installation in Tri Alpha Energy [61]. This will allow a better understanding of transport and turbulent processes in FRC, especially in collisional plasma, and explain the contradictions between theory and experiment [27, 62, 63].

The parameters and operating modes of D–³He reactors based on the existing magnetic confinement systems: tokamak, spherical tokamak, stellarator, reversed magnetic configuration or an elongated compact torus are investigated. The analysis shows that the D–³He FRC thermonuclear power plant (TNPP) should have the cost of electricity, including ³He production on the Moon [35, 64, 65], lower than the price in the D–T FRC TNPP [42, 66]. Moreover, low neutron loads, high specific power density and the absence of a blanket for tritium reproduction greatly simplify the engineering design of the D–³He FRC and make its development much simpler than D–T commercial reactors [10]. Open magnetic field lines near the separatrix make it possible to use direct conversion of the thermonuclear energy of the reaction products to electricity with high efficiency, thereby reducing the outlived and lost power needed to generate each kilowatt of electrical energy. Thus, the D–³He FRC TNPP can achieve sufficient commercial attractiveness in the global energy market. This alternative direction for energy production will be able to replace exhaustible resources of fossil fuels.

1.5. Spherical tokamak reactor on D–³He fuel

The analysis of the prospects for creating a low-radioactive reactor for the D–³He mixture is advisable to start with a tokamak, since the efforts of the international thermonuclear program are directed to this system. The calculation method ITER Physics Basis (IPB), which is the basis of the ITER project [2, 67], allows predicting the parameters of the tokamak reactor with a high degree of reliability. At low β, typical for a classical tokamak with an aspect ratio of $A \approx 3$, the use of D–³He fuel is ineffective [28]. One of the reasons that a D–³He reactor requires a confinement system with $\beta \sim 1$ is that at the same plasma pressures, the rate of the D–³He reaction is many times lower than the rate of the D–T reaction. The confinement of a plasma with $\beta \sim 1$ is possible in spherical tokamaks with a low aspect ratio $A = 1.1$–2 [68–72]. Therefore, consider a spherical tokamak with $\beta \approx 0.5$, for which the basic regularities of IPB are satisfied [69].

Today the tokamaks occupy a leading position among magnetic traps for plasma confinement both in terms of achieved parameters, and in terms of research and financial costs. The physical processes in the tokamak plasma have been fairly well studied in experiments, therefore the results of the analysis of the possible parameters of the reactor based on tokamak and its varieties are most credible. If we take into account that the physical processes in the plasma of a spherical tokamak are also subject to IPB regularities, then the spherical tokamak looks, at first glance, almost an ideal candidate for the D–^{3}He reactor. An important technical task in developing the concept of such a reactor is to justify the possibility of creating high magnetic fields by superconducting magnetic coils under the conditions of the chosen geometry.

The induction of the magnetic field, the size and power of a competitive thermonuclear reactor should be such that the thermonuclear power released per unit volume of plasma is not less than 2 MW/m^3. Analyzing the results of Refs. [28, 29, 73, 74], on can say that this condition is not satisfied in the D–^{3}He reactor with a vacuum toroidal field on the magnetic axis B_0 = 2–3 T. To create a compact and simultaneously powerful reactor, $B_0 \approx 5$ T is required. The possibility of creating such a reactor was analyzed in [33]. In the corresponding integrated model, the main parameters of the magnetic configuration and the plasma confinement time are calculated using the IPB technique with modification for low aspect ratios. The plasma energy balance is also considered taking into account the features of the D–^{3}He plasma.

As a deuterium–tritium reactor, the use of a spherical tokamak seems problematic, since its compact geometry makes it difficult to locate the neutron shield and blanket for the reproduction of tritium. In the case of the D–^{3}He fuel reactor, the neutron shielding is much smaller, and the blanket is no longer needed. Here we consider the D–^{3}He reactor without the production of helium-3.

The required high power level in a D–^{3}He reactor based on a spherical tokamak can be achieved if the magnetic field is increased to the theoretical limit for this system, taking into account the capabilities of the existing superconducting materials. The maximum value of magnetic field induction on the surface of superconducting windings should not exceed the critical value B_c for the type of superconducting material used.

Parametric analysis of the spherical tokamak reactor on D–^{3}He fuel was carried out in [73]. An analysis of the energy balance of a spherical tokamak reactor on a D–^{3}Hefuel is given in [28, 29, 74]. In [75], pressure and current profiles were analyzed that corresponded to different models of transport in a D–^{3}He plasma of a spherical

Table 12. Parameters of D-³He reactors based on spherical tokamak [29, 74, 75]

Parameter	[74]	[75]	[29], var. 1	[29], var. 2
Minor radius of the plasma column a, m	6.15	3.5	3	2
Major radius of the plasma column R, m	8	6.2	4.5	3
Induction of the vacuum magnetic field on the magnetic axis B_0, Tl	2.7	4.2	3	5
Current in plasma I_p, MA	128	49	200	140
The average ratio of plasma pressure and the magnetic field $\langle \beta \rangle$	0.32	0.52	0.5	0.4
The average temperature of the plasma $\langle T \rangle$, keV	43	35	40	40
Fusion power P_{fus}, MW	6100	3440	1500	1500

tokamak. Table 12 shows the parameters of the reactors considered in [29, 74, 75].

The dimensions of the reactor considered in [74] look excessively large, and the thermonuclear power per unit volume is only 0.34 MW/m³. In the first variant, considered in [29], the magnetic field is low. In the second variant, the field is larger and the sizes smaller. The thermonuclear power in both of these variants is the same and is equal to 1500 MW. Taking the efficiency of the system for converting the energy released into electricity to 40%, we find that the electric power of such a reactor is 600 MW. The second of these options is preferable, since it corresponds to a more compact system, and the energy release density reaches about 1.7 MW/m³.

The geometry of the plasma in a spherical tokamak (see Fig. 1.15) is characterized by a minor radius a, a major radius R, the aspect ratio $A = R/a$, the elongation of the section k and the triangularity parameter δ. Outside the plasma is the protection of the superconducting coils from neutron fluxes from the plasma. The coils are closed to the central rod, which is the most important element in the design of the magnetic system. Due to the compactness of the system, the radius of the central rod r_0 can not be too large. At the same time, the largest current flowing through the central rod is formed from currents flowing in toroidal coils. Therefore, the size of r_0 can not be too small.

The radius of the central rod is $r_0 = R - a - \Delta_s - \Delta_0$, where Δ_s is the thickness of the neutron shield (together with the first wall), Δ_0 is the gap between the plasma and the first wall. The magnetic

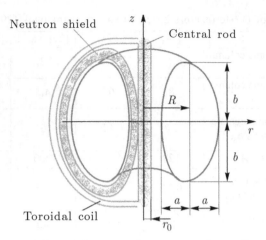

Fig. 1.15. Simplified geometric diagram of plasma in a spherical tokamak (the area of the divertor is not shown).

field is maximal on the surface of the central rod. The ratio of the maximum field to the field on the magnetic axis (in vacuum)

$$\frac{B_{max}}{B_0} = \frac{B(r_0)}{B(R)} = \frac{R}{R - a - \Delta_s - \Delta_0}. \tag{1.12}$$

Since the power of the reactor is proportional to the square of the plasma density, and the density is proportional to the square of the magnetic induction, the value of B_0 should be as large as possible. On the other hand, it is limited by the capabilities of modern superconducting materials. It is desirable to choose the smallest possible aspect ratio, taking into account the dimensions of the central rod and neutron shielding. Based on these considerations, several options were analyzed and the following values were accepted: $B_0 = 5.5$ T, $a = 2$ m, $A = 1.7$, $\Delta_s = 0.3$ m, $\Delta_0 = 0.15$ m. For the considered variant, $r_0 = 0.95$ m, $B_{max} = B(r_0) = 19.7$ T.

We also take into account that the current in the plasma creates a poloidal magnetic field B_p, perpendicular to the toroidal field. In spherical tokamaks on the surface of a plasma column $B_p \approx B_0$. Consequently, the maximum field $B_{max} = \sqrt{[B(r_0)]^2 + B_p^2} \approx 20.5$ T. Such a value of magnetic induction is permissible for the existing superconducting materials. For example, for a superconductor based on the niobium–tin (Nb_3Sn) intermetallic, the critical field at a liquid helium temperature of 4.2 K is B_c = 24.5 T, and the maximum the current density at $B \approx 20$ T is $j_{max} \approx 0.4 \cdot 10^9$ A/m^2.

The current strength in the central rod is $I \approx 2\pi r_0 B(r_0)/\mu_0$, where μ_0 is the magnetic constant. For the selected set of parameters one obtains $I = 9.35 \cdot 10^7$ A. The cross-sectional area of the superconductor is 20.28 m^2 (10% coverage). Then the current density is $j = 0.33 \cdot 10^9$ A/m^2. This value is within the limits of admissible values.

The considered variant of the magnetic system of a spherical tokamak reactor can be considered as the case of limiting possibilities of superconducting materials.

The basic calculated dependences of ITER Physics Basis for a tokamak reactor are presented in a book by K. Miyamoto [76, § 16.11], which also shows the calculation sequence and results for the ITER-FEAT variant [67, 77].

From the point of view of the calculation method, two differences of the reactor we are considering on the basis of the spherical tokamak with D–³He fuel from ITER, which is based on the classical tokamak with D–T fuel, are significant. First, the aspect ratio A and the corresponding achievable values of the parameter β: in ITER $A > 3$, $\beta < 0.05$; in our case $A < 2$, $\beta \sim 0.5$. Secondly, the plasma temperature in the D–³He reactor ($T \approx 50$–70 keV) is substantially higher than in the D–T reactor. Therefore, for our research, a technique was specially developed [33], combining the dependences of the ITER Physics Basis with the dependences specific for the spherical tokamaks.

In the calculations, the sizes and shape of the plasma were set; induction of the toroidal magnetic field on the B_0 axis; temperature T_0 and the value of β_0 in the centre of the plasma column; the relative content of impurities in the plasma; the current in the plasma I_p and the value of the power gain factor in the plasma Q. The correspondence between the values of the parameters for the permissible ranges for spherical tokamaks was analyzed. The plasma energy confinement time was determined from the energy balance and compared with the value predicted by the IPB98(y, 2) dependence for the ITER. The density of the energy flux of neutrons from the plasma was limited to 0.25 MW.

Calculation of the geometry of the cross section and volume of the plasma in a tokamak was carried out according to the method described in [76]. The maximum elongation of the plasma cross section is $k = 3.7$, as, for example, in the conceptual design of

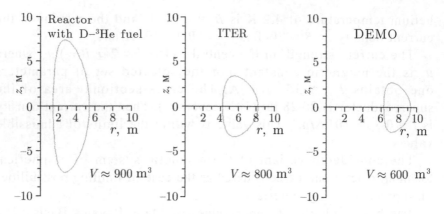

Fig. 1.16. The shape of the plasma column cross section in a D–³He reactor based on a spherical tokamak, ITER and DEMO.

the spherical ARIES (D–T fuel) tokamak [78]. The triangularity parameter of the section is $\delta = 0.35$ (approximately as in ITER). In Fig. 1.16 a comparison is made of the cross-sectional shape of the D–³He reactor under consideration, ITER and DEMO (Russian project [78]). The plasma volume V for these reactors is approximately 900 m³, 800 m³ and 600 m³. The safety factor is [74, 49]

$$q_a = \frac{5aB_0}{2AI_p} \frac{[1+k^2(1+2\delta^2-1.22\delta^3)](1.17-0.65A^{-1})}{(1-A^{-2})^2}. \qquad (1.13)$$

Here, the current in the plasma I_p is measured in MA.

The distributions of density, temperature, and pressure for all components of the plasma are given in the form

$$n = n_0(1-\rho^2)^{\alpha_n}, \qquad (1.14)$$

$$T = T_0(1-\rho^2)^{\alpha_T}, \qquad (1.15)$$

$$p = p_0(1-\rho^2)^{(\alpha_n+\alpha_T)}, \qquad (1.16)$$

where n_0, T_0 and p_0 are the values of density, temperature and pressure on the axis of the plasma column, ρ is the normalized radius.

The pressure on the axis of the plasma column is $p_0 = \beta_0 \dfrac{B_0^2}{2\mu_0}$, where β_0 is the value of the parameter β on the axis. The averages are related to the values on the axis by the relations: $\langle n \rangle =$

$n_0/(1 + \alpha_n)$, $\langle T \rangle = T_0/(1 + \alpha_T)$, $\langle p \rangle = p_0/(1 + \alpha_n + \alpha_T)$, $\langle \beta \rangle = \beta_0/(1 + \alpha_n + \alpha_T)$, where the angle brackets mean averaging over the plasma volume.

The quantity $\langle \beta \rangle$ satisfies relation

$$\langle \beta \rangle = 0.01 \beta_N \frac{I_p}{aB_0}, \tag{1.17}$$

where β_N is the normalized beta (the Troyon coefficient).

The temperatures of ions and electrons are assumed to be equal: $T_e = T_i = T$. When calculating the density of plasma components, the quasi-neutrality condition is taken into account

$$n_e = \sum_i Z_i n_i, \tag{1.18}$$

where n_e is the electron density, n_i is the ion concentration, Z_i is the ion charge, the summation is over all ion types.

The average electron density must satisfy the condition imposed on the Greenwald number

$$N_G = \frac{\pi a^2 \langle n_e \rangle}{10^{20} I_p} \leqslant 1. \tag{1.19}$$

The thermal energy of the plasma is

$$E_{th} = \frac{3}{2} \left\langle \sum_i n_i k_B T_i + n_e k_B T_e \right\rangle V = \frac{3}{2} \langle \beta \rangle \frac{B_0^2}{2\mu_0} V, \tag{1.20}$$

where V is the volume of the plasma.

The steady-state energy balance of a thermonuclear plasma is

$$(1 - f_{fast})(P_{fus} - P_n) + P_{aux} = P_r + \frac{E_{th}}{\tau_E}. \tag{1.21}$$

Here, on the left side, are the power of the plasma heating sources, in the right part – the power of the losses.

In the charged particles that are confined by a magnetic field, a power equal to the difference in the total fusion power P_{fus} and the neutron power P_n are released (the neutrons instantly leave the plasma volume). Let us take the share of losses of fast charged particle–reaction products $f_{fast} = 0.05$ (as in ITER). Also, to maintain the predetermined mode, the P_{aux} power is supplied from the external heating sources to the plasma. The losses include the radiation power P_r (radiative losses) and the transfer of thermal energy, characterized by the energy confinement time τ_E.

The most important indicator of the efficiency of energy production in a thermonuclear reactor is the power amplification factor in the plasma

$$Q = P_{fus} / P_{aux}. \tag{1.22}$$

In this study, the value of Q is given in accordance with the requirements of the energy efficiency of the system. Thus, for an industrial fusion reactor $Q \geq 10$.

The power of radiation losses is

$$P_r = P_b + P_s, \tag{1.23}$$

where P_b is the bremsstrahlung power, P_s is the cyclotron radiation power.

Generally speaking, the radiation losses should also include the power of line radiation. In our case, we confined ourselves to considering only bremsstrahlung and cyclotron radiation, other types of radiation are neglected, since the first two, in our opinion, are dominant under the conditions of the D–^{3}He reactor. We note that the power of line radiation in the ITER-FEAT is comparable to the bremsstrahlung power. Line radiation is largely due to the presence of heavy impurities such as, for example, argon, which in ITER is used for thermal protection of the divertor. Since for D–^{3}He plasmas contamination with heavy impurities is highly undesirable (due to a large fraction of the radiation losses in the energy balance), then the option with lithium coating of the components facing the plasma is preferable. Lithium is the material with the lowest nuclear charge. The advantages of liquid lithium in combination with a capillary–porous metal substrate were demonstrated in experiments on lithium tokamaks [80].

When analyzing the energy balance, the rates of thermonuclear reactions are calculated from the formulas in [81]. For bremsstrahlung, relations are used with allowance for relativistic effects [82]. Synchrotron losses are calculated according to [83].

In our calculation algorithm, the value of the power gain Q is set, which for known P_{fus} allows us to determine the value of P_{aux}. In the balance equation, the quantities P_{fus}, P_n, P_r and E_{th} are determined by the composition of the plasma and the temperature, the value of f_{fast} is given. To calculate the cyclotron radiation power P_s entering in P_r, the reflection coefficient of cyclotron radiation by the wall R_w is given. From the energy balance, the required energy confinement time τ_E can be determined.

The energy confinement time τ_E, found from the energy balance, must be compared with what is predicted by the dependence obtained as a result of processing a large volume of experimental data obtained at many tokamaks. To do this, we use the dependence IPB98(y, 2) [2]

$$\tau_E^{98y2} = 0.0562 I_p^{0.93} B_0^{0.15} M^{0.19} (\langle n_e \rangle / 10^{19})^{0.41} a^{1.97} A^{1.39} k^{0.78} P_L^{-0.69}, \qquad (1.24)$$

where M is the average mass number of ions, P_L is the loss power (in MW) due to transfers, equal to the required absorbed heating power (external and reaction products) minus the radiation losses.

As one of the optimization criteria, the confinement enhancement factor:

$$H_{y2} = \tau_E / \tau_E^{98y2}. \qquad (1.25)$$

Acceptable is the range of values H_{y2} = 1.2–1.5, which corresponds to modern experiments on spherical tokamaks [69, 71, 72].

For calculations it is necessary to set the temperature at the center of the plasma column T_0, the ratio between the fuel components $x_{3He} = n_{3He}/n_D$ (n_{3He} is the concentration of helium-3, n_D is the deuterium concentration) and the relative impurity content $x_{imp} = n_{imp}/n_D$ (n_{imp} is the impurity concentration). The values of T_0 and x_{3He} are determined as a result of optimization. The main criterion of optimization is factor H_{y2}, which should be minimal. An additional optimization criterion for the quantity x_{3He} is the neutron flux density from the plasma $J_n = P_n/S_0$, where S_0 is the area, which in this case was assumed equal to the surface area of the plasma.

Let's consider some variants. In the progressive version, a high elongation of the cross section is assumed to be k = 3.7 (as, for example, in the ARIES-ST project [84]) and a high normalized beta, β_N = 5. The average energy release capacity in such a reactor can be 3.5 MW/m³.

Figures 1.17–1.19 present the results of optimization. According to calculations, the optimum temperature at the centre of the plasma is T_0 = 62 keV in the regime with Q = 10. From the point of view of

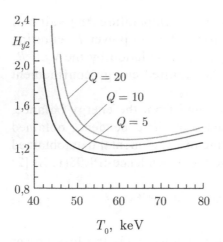

Fig. 1.17. The confinement enhancement H_{y2} as a function of temperature at the centre of the plasma column at $x_{3_{He}} = 1$.

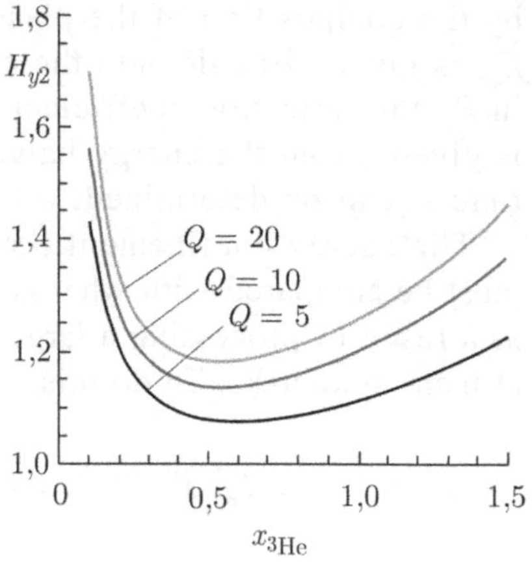

Fig. 1.18. Factor H_{y2} as a function of the relative concentration of helium-3 at $T_0 = 62$ keV.

the value of H_{y2}, the optimum is the ratio $x_{3_{He}} = 0.5$–0.6. In this case, the neutron flux is approximately twice as high as at $x_{3_{He}} = 1$, and the difference of H_{y2} is only about 10%. Therefore, we take $x_{3_{He}} = 1$. Lithium ions Li^{3+}, beryllium Be^{4+} and argon Ar^{18+} were considered as possible impurities. Beryllium and argon are impurities in the plasma of ITER. Lithium is a promising coating material with the smallest possible ion charge. For admixtures, the maximum permissible concentrations were determined, the calculation results of which are shown in Fig. 1.20.

The limiting concentration was taken to be the value corresponding to a sharp increase in H_{y2} (see Fig, 1.20). For lithium, the limiting content is $x_{Li} \approx 0.2$, for beryllium $x_{Be} \approx 0.1$, the presence of even a small amount of argon is undesirable.

Thermonuclear reactions are also a source of impurities. Such products of the D–D reaction, such as helium-3 and tritium, react. Products of the D–^{3}He reaction (protons and alpha particles) do not burn. The content of such products in plasma is determined from the balance of their birth rates and losses. The characteristic loss times for the alpha particles τ_α and protons τ_p are specified.

The results of calculations of three variants of the D–^{3}He reactor are presented in Table 13. The ITER-FEAT parameters and test calculation results are also presented there. More detailed data on the modes of the spherical tokamak-reactor are contained in [33].

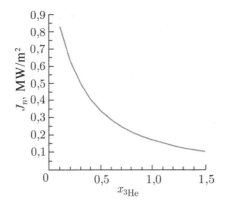

Fig. 1.19. The energy flux of neutrons from the plasma as a function of the relative concentration of helium-3 at $T_0 = 62$ keV.

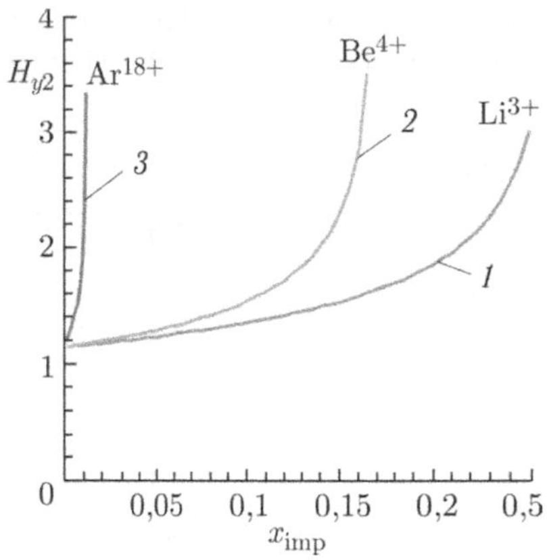

Fig. 1.20. The H_{y2} factor depends on the relative concentrations of impurities: 1 – Li^{3+} ($x_{Be} = x_{Ar} = 0$), $2 - Be^{4+}$ ($x_{Li} = x_{Ar} = 0$), $3 - Ar^{18+}$ ($x_{Be} = 0.025$, $x_{Li} = 0$).

The purpose of the test calculation was to check the calculation method. As can be seen, the values determined by calculation (in the test version they are marked with *) are in good agreement with the ITER-FEAT parameters.

In addition to the scaling of IPB98(y, 2), we can also consider the scaling of Valovic et al. [85], which is adequate to the experimental results on spherical MAST and NSTX tokamaks. Since this scaling in its original form does not take into account a number of parameters, its generalization proposed in [86],

$$\tau_E^{Val} = 0.252 I_p^{0.59} B_0^{1.4} P_L^{-0.73} M^{0.19} R^{1.97} (1/A)^{0.58} k^{0.78}. \qquad (1.26)$$

Despite the difference (1.24) and (1.26), the calculation yields close numerical confinement times.

Table 13 shows: a minor radius a; a major radius R; aspect ratio $A = R/a$; elongation k; the triangularity parameter δ; magnetic induction on the magnetic axis in vacuum B_0; current in plasma I_p; safety factor q_{95} on the surface, covering 95% of the magnetic flux; average beta $\langle \beta \rangle$; normalized beta β_N; the Greenwald number N_G; average electron density $\langle n_e \rangle$; average ion and electron temperatures $\langle T_i \rangle$ and $\langle T_e \rangle$; the temperature and density indexes α_T and α_n; coefficient of wall reflection of cyclotron radiation R_w; thermal energy of the plasma E_{th}; fusion power P_{fus}; relations P_n/P_{fus}, P_b/P_{fus}, P_s/P_{fus}, P_r/P_{fus}; the plasma gain Q; energy confinement time τ_E;

Alternative Fusion Fuels and Systems

Table 13. Parameters of the spherical tokamak reactor with D-^{3}He-fuel (variants ST-1, ST-2 and ST-3), the ITER-FEAT project [67] and the results of the test calculation

Parameter	D-^{3}He ST-1	D-^{3}He ST-2	D-^{3}He ST-3	ITER–FEAT	Test calculation
a, m	2.0	2.0	2.0	2.0	2.0
R, m	3.4	3.4	3.4	6.2	6.2
$A = R/a$	1.7	1.7	1.7	3.1	3.1
k	3.7	2.8	2.8	1.7	1.7
δ	0.35	0.5	0.5	0.35	0.35
B_0, T	5.5	5.5	5.5	5.3	5.3
I_p, MA	110	110	110	15	15
q_{95}	4.7	3.1	3.1	3.0	3.0*
$\langle \beta \rangle$	0.5	0.5	0.25	0.025	0.025
β_N	5.0	5.0	2.5	1.77	1.77*
N_G	0.56	0.56	0.28	0.85	0.83*
$\langle \vartheta n_e \rangle$, 10^{20} m^{-3}	4.92	4.92	2.45	1.01	1.03*
$\langle T_i \rangle$, keV	44.3	44.3	44.3	8.1	8.1
$\langle T_e \rangle$, keV	44.3	44.3	44.3	8.9	8.9
α_T/α_n	0.4/0.2	0.4/0.2	0.4/0.2	1.0/0.1	1.0/0.1
R_w	0.85	0.65	0.85	—	0.5
E_{th}, MJ	8124	5900	2950	325	324*
P_{fus}, MW	3064	2225	582	410	410*
P_n/P_{fus}	0.053	0.053	0.064	0.8	0.8*
P_b/P_{fus}	0.57	0.57	0.53	0.057	0.077*
P_s/P_{fus}	0.089	0.145	0.26	0.0195	0.0195*
P_r/P_{fus}	0.67	0.71	0.79	0.117	0.097*
Q	10	10	10	10	10
τ_E, s	7.8	9.4	25.7	3.7	3.7*
$H_{y2} = \tau_E / \tau_E^{98y2}$	1.24	1.31	1.48	1.0	1.0*
$H_{Val} = \tau_E / \tau_E^{Val}$	1.06	1.09	0.87	—	—
J_n, MW/m^2	0.16	0.16	0.05	0.4	0.4*

confinement enhancement factor $H_{y2} = \tau_E / \tau_E^{98y2}$; factor $H_{Val} = \tau_E / \tau_E^{Val}$; the average energy flux of neutrons from the plasma J_n.

In the progressive version (ST-1), the thermonuclear power $P_{fus} \approx 3000$ MW is high enough for an industrial fusion reactor. The electric power of such a reactor can exceed 1000 MW. The share of the bootstrap current estimated by [87] is approximately 95% in this version.

From a physical point of view, the most serious difficulties in implementing a progressive variant are associated with high elongation values $k = b/a = 3.7$ and normalized beta $\beta_N = 5$. In today's experiments, the maximum elongation is $k = 2.8$. With a decrease in the elongation to this value and a corresponding decrease in the plasma volume, the thermonuclear power is reduced to 2200 MW (option ST-2). This value can be considered acceptable. However, a decrease in the plasma cross section size may be undesirable from the point of view of maintaining a large current in the plasma.

The value of $\beta_N = 5$ is typical for modern spherical tokamaks with $B_0 \approx 1$ T. At $B_0 \approx 5$ T, the limiting value of β_N can be significantly smaller [88]. A decrease in this parameter will lead to a decrease in the pressure and density of the plasma and, as a consequence, to a drop in the power of the reactor. In the conservative variant (ST-3), $k = 2.8$, $\beta_N = 2.5$. The thermonuclear power of such a reactor is unacceptably low. Thus, the prospects of the considered concept of the reactor depend on the physical possibilities of realizing regimes with high values of k and β_N for a relatively strong magnetic field. The corresponding regimes suggest significant progress in comparison with the limiting parameters of today's spherical tokamaks [89]. As far as this progress is possible, only future experiments can show.

The advantage of the D–³He reactor lies primarily in low neutron activity, which significantly alleviates the problem of the first wall. The need to develop a reasonable concept of such a reactor is associated with the possibility of extracting the ³He isotope from the lunar soil.

Sources of thermonuclear neutrons with magnetic confinement

2.1. Source of neutrons based on tokamak

Today, tokamaks are the most promising prospects for the development of an industrial thermonuclear reactor. It is on this system that the design of the experimental reactor ITER [2] is based. Modern tokamaks approach the conditions for the realization of regimes in which, when operating on a deuterium–tritium (D–T) mixture, the energy yield is approximately equal to the power input to the plasma, i.e., the power amplification factor in the plasma is $Q \approx 1$. However, from the creation of a thermonuclear power station this direction is separated by a very serious problem of the radiation resistance of components facing the plasma in the conditions of neutron fluxes of 1–3 MW/m². Perhaps a more rapid introduction to power engineering will be found by less powerful devices with $Q \approx 1$. Thermonuclear systems with $Q \approx 1$ scales of the already existing experimental facilities can be used in power engineering as neutron sources for hybrid fusion reactors with a subcritical blanket. The actuality of such reactors is associated with the possibility of transmutation ^{238}U, ^{232}Th raw material isotopes with thermonuclear neutrons with the release of energy and the production of fissile isotopes ^{239}Pu, ^{233}U. The latter are suitable for use as a fuel for nuclear reactors on thermal neutrons. The energy released in the blanket can be tens of times greater than the energy of neutrons released in the thermonuclear plasma. The total gain in the hybrid reactor $Q_{\text{tot}} \approx 10$ and more.

When a plasma is heated by a beam of fast particles, a significant population of fast ions is formed. The rate of a thermonuclear reaction involving fast ions is much higher than the reaction rate in a Maxwellian plasma. Consequently, the requirements to the size of the reactor and the plasma energy confinement time are reduced. Taking into account the reduced requirements, the creation of neutron sources is possible not only on the basis of tokamaks, but also on the basis of magnetic traps simpler from the technical point of view.

Calculation of the plasma parameters in the tokamak reactor and the energy balance is performed according to the previously developed model [33], verified on the ITER modes. The energy balance in a plasma with injection heating is expressed by the equation

$$(1 - f_{fast})(P_{fus} - P_n + P_{inj}) = P_{rad} + \frac{E_{th}}{\tau_E}, \qquad (2.1)$$

where P_{fus} is the thermonuclear power, P_n is the power in the neutrons, P_{inj} is the absorbed injection power, P_{rad} is the radiation power, E_{th} is the thermal energy of the plasma, τ_E is the energy confinement time, f_{fast} is the fraction of the energy loss of the fast particles ($f_{fast} = 0.05$).

The plasma gain is determined as follows:

$$Q = P_{fus} / P_{inj}. \qquad (2.2)$$

In the calculation algorithm, the required confinement time τ_E of the thermal components is determined from the balance equation (2.1). The obtained value is compared with the scaling value of IPB98y2 [2]. The equality of the indicated values of the confinement time is achieved as a result of iterations.

With a sufficiently large number of beams of fast atoms and a relatively small radius of the plasma column, the temperature and density can be practically constant over the plasma cross section. In the present study, we take the temperature and density to be homogeneous, which corresponds to the zero-dimensional approximation. Consideration of any other spatial distributions of temperature and density within the framework of our study seems less valid, since they require additional information on the energy release on trajectories of the beams, which is not included in the range of issues under consideration. To justify that the adopted approximation does not lead to an overestimation of the plasma gain Q, a special calculation of the parameters of the zero-dimensional analog of ITER (ITER-0D) is performed.

The rate of the thermonuclear reaction is

$$\gamma = n_D n_T \langle \sigma v \rangle, \tag{2.3}$$

where n_D and n_T are the concentrations of deuterium and tritium nuclei, $\langle \sigma v \rangle$ is the reaction rate parameter, σ is the reaction cross section, and v is the velocity of the colliding particles; Angular brackets mean averaging over the velocity distribution functions of particles.

In the Maxwellian plasma, the reaction rate parameter $\langle \sigma v \rangle$ depends on the temperature T_i of the reacting components (deuterium and tritium ions). The electron temperature T_e may differ slightly from the ion temperature (depending on the heating conditions). Here, to simplify the analysis, we adopted $T_e = T_i = T$.

For the cross section of the reaction, we use approximating relationships [81]. The cross section for the D–T reaction reaches a maximum at the energy of the colliding nuclei $E = 100$–200 keV. As the energy decreases, the cross section drops sharply.

For the reaction rate $\langle \sigma v \rangle$ in the Maxwellian plasma one can use the formulas [81]. In the conditions under consideration, the presence of fast particles must be taken into account. The velocity distribution function of fast particles can be found as a result of a numerical solution of the Fokker–Planck equation. The slowing down of fast particles is well described by the classical theory of Coulomb collisions. Approximate solutions are known for this case [90]. The beam relaxation time is approximately equal to the slow-down time τ_s.

In the present study, in order to take into account the beam velocity, the approximation of the 'shifted' Maxwellian distribution is used [91]. It is assumed that deuterium atoms are injected. Injection replenishes both energy loss and loss of deuterium. It is assumed that the losses of tritium are replenished by the injection of pellets.

When the plasma components move relative to one another with a high velocity V, the velocity of the colliding particles on average becomes higher by the relative velocity of the components. This increases the reaction rate parameter in comparison with the Maxwellian plasma. The use of the 'shifted' Maxwellian velocity distribution function in calculating the reaction rate parameter leads to the expression [90]

$$\langle \sigma v \rangle = \frac{2}{V} \sqrt{\frac{M}{2\pi k T_i}} \int_0^\infty \mathrm{sh}\left(\frac{MVv}{kT_i}\right) \exp\left(-\frac{M(v^2 + V^2)}{2kT_i}\right) v^2 \sigma(v) dv, \qquad (2.4)$$

where k is the Boltzmann constant, M is the reduced mass of the colliding particles.

The results of calculations are presented in Table 14. As the calculation of the zero-dimensional analog (ITER-0D) has shown, in comparison with ITER, in which the spatial distributions of density are inhomogeneous in volume, the gain is reduced to $Q = 5$. This is because in the zero-dimensional analogue for the same the value of the average relative pressure of the plasma $\langle \beta \rangle$, the average plasma density and thermonuclear power are lower than in the case of spatially inhomogeneous temperature and density profiles. The resulting decrease in Q means that using the zero-dimensional approximation does not lead to unjustifiably optimistic results.

Both for the zero-dimensional analogue and for the neutron source, the values of the magnetic configuration corresponding to ITER are taken: the elongation cross section is $k = 1.7$, the triangularity factor of the cross section is $\delta = 0.35$, the safety factor at the plasma boundary is $q_a = 3$. The energy of the injected atoms is assumed equal to 80 keV. As can be seen from Table 14, the slow-down time of the beam in the modes close to ITER is much less than the confinement time. With such rapid relaxation, the effect of increasing the reaction rate is practically invisible.

Under the conditions of a neutron source, the slow-down time is much longer than the confinement time, which makes it possible to calculate the realization of the plasma-beam regime of a thermonuclear reaction with an increased velocity. The estimated plasma dimensions of such a source of neutrons correspond to the level of today's experimental installations. With a thermonuclear power of 100 MW, the power in the blanket can be ~3 GW. Such a system fully corresponds to modern power reactors with an electric power of ~1 GW.

We also note that in the transition from homogeneous temperature and plasma density, some improvement in performance is possible due to the formation of optimal profiles.

Table 14. ITER parameters, calculation results of a zero-dimensional analog and a source of thermonuclear neutrons with injection (E_0 = 80 keV injection energy)

Parameter	ITER-FEAT [2]	Zero-dimensional analogue	Neutron source
Minor plasma radius a, m	2.0	2.0	0.85
Major radius R, m	6.2	6.2	4.3
Aspect ratio $A = R/a$	3.1	3.1	5.1
Plasma volume V, m³	828	828	104
Induction on the axis B_0, T	5.3	5.3	4.5
Current in plasma I_p, MA	15	15	3.1
Average Beta $\langle \beta \rangle$	0.025	0.025	0.015
Normalized beta (Troyon number) β_N	1.77	1.77	1.83
Greenwald number N_G	0.85	0.78	0.27
The average electron density $\langle n_e \rangle$, 10^{20} m⁻³	1.01	0.85	0.37
Fusion power P_{fus}, MW	410	310	100
Gain $Q = P_{fus}/P_{inj}$	10	5	1
Energy confinement time τ_E, s	3.7	2.6	0.16
Beam slow-down time τ_s, s	—	1.2	2.9

2.2. Open trap systems

The largest experimental installation of the class of open traps in Russia today is the gas-dynamic trap (GDT) in the G.I. Budker Institute of Nuclear Physics of the Siberian Branch of the Russian Academy of Sciences in Novosibirsk. This is a classic mirror with a large mirror ratio. Figure 2.1 shows the main nodes and elements of the GDT, which is an axially symmetric magnetic system for confining the plasma. To conduct studies of a compact anisotropic plasmoid, an additional compact mirror cell (CMS) with a field of 2.5 T and a mirror ratio of ~2 was added to it. A warm plasma with a temperature of 100 eV and a density of 10^{19} m⁻³ flows into the CMC from the main part of the GDT. Focused atomic beams with a particle energy of 25 keV and a power of 1 MW. As a result of ionization of the beams, a cluster of fast ions with an average energy of 15 keV

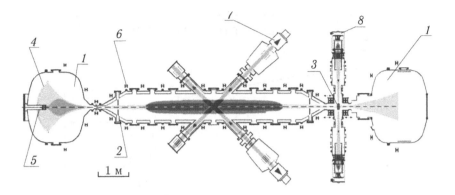

Fig. 2.1. Scheme of experiment with a compact mirror in the GDT installation:
1 – end tanks, *2* – central section, *3* – compact mirror cell, *4* – plasma receiver,
5 – plasma source, *6* – coils of the magnetic system, *7* – heating injectors for atoms,
8 – injectors for CMC.

and a density several times higher than the density of the flowing
plasma should form in the CMC. The geometry of the magnetic field
of the GDT and CMC is shown in Fig. 2.2.

The magnetic field, taking into account the plasma diamagnetism
in the open trap, can be represented by the dependence

$$B_p = B \cdot \sqrt{1 - \beta}, \tag{2.5}$$

where B_p is the field in the plasma, B is the vacuum magnetic field,
and β is the average beta.

The β (beta) ratio is one of the main characteristics for magnetic
plasma confinement systems and determines the operating modes
of many installations, especially when it comes to thermonuclear
perspectives of the system and primarily as a power plant. The
maximum value of energy content in fast ions corresponds to the
maximum value of β. And it, as well as the energy confinement
time, depends on the magnitude of the external magnetic field and
temperature that enter the ion gyroradius $\rho_i = 0.00456\sqrt{T_i m_i / Z_i B}$,
where m_i is the mass of the ion in atomic mass units, Z_i is the ion
charge.

The characteristic energy lifetime of fast ions in the GDT is
determined mainly by their Coulomb slow-down on the electrons
and is expressed by the equation:

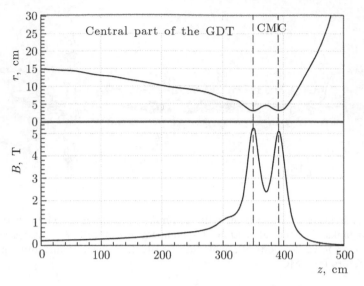

Fig. 2.2. The configuration of the magnetic field in the GDT (right side): the line of force r_0 = 15 cm (top) and the distribution of the magnetic field (bottom) along the axis of the GDT and the compact mirror cell.

$$\tau_E^{GDT} = \frac{3m_i T_e^{3/2}}{8\sqrt{2\pi m_e} e^4 \lambda n_e},$$

(2.6)

where m_i and m_e are the ion and electron masses, respectively, T_e and n_e are the electron temperature and density, λ is the Coulomb logarithm.

The high energy content of fast ions in the GDT setup corresponds to the following beta parameters: 600 J – $\beta \approx 0.25$, 1200 J – $\beta \approx 0.5$. Depending on the mode of operation, the purpose and possibilities of using compact systems are subdivided.

Due to its advantages, first of all simple geometry, GDT can be considered as a prototype of the source of thermonuclear neutrons for materials science, utilization (afterburning) of minor actinides and, possibly, hybrid systems 'fusion–fission'.

The thermonuclear regimes of a quasistationary axially symmetric open trap, for which the longitudinal losses are determining, were considered in Refs. [92, 93]. The peculiarity of the regimes considered is the powerful injection of fast particles into the plasma. As shown by the simulation, the modes with a power gain factor in the plasma $Q_{pl} \approx 1$ (Q_{pl} is the ratio of the thermonuclear power to the absorbed power of external heating) can be realized in a relatively compact system. Such a system is comparable in size to the existing experimental installations of the open type, such as, for example, the GDT – gas-dynamic trap [94]. Note that powerful injection in such systems contributes to the formation of positive potential barriers at the ends of the open system, which in turn improves longitudinal confinement.

Further, an analysis is made of the possible application of a thermonuclear system based on an axially symmetric open trap. From the technical point of view, the essential advantage of such a trap is the simplicity of the design. In addition, axial symmetry removes the problem of neoclassical transport. Under the condition of stabilization of the magnetohydrodynamic and kinetic instabilities characteristic of open traps, the efficiency of confinement of particles and energy in such a trap is determined by the classical longitudinal losses. In this work, we consider the regimes for the energy balance of which the longitudinal losses are determining. The transverse turbulent transport in the model used is set by some characteristic time exceeding the longitudinal loss time. In the questions of stabilization of instabilities, we confine ourselves to an analysis of the existing theoretical and experimental data applicable to an axially symmetric system.

It is not easy to imagine a reliable extrapolation of parameters from today's level of open traps to the region of thermonuclear reactor regimes with a power gain factor in the plasma $Q > 10$. Therefore, in this study, we focused on comparatively low values of $Q \approx 1$, which may be sufficient for the source of thermonuclear neutrons of the hybrid reactor.

The main goal of this chapter is to analyze the conditions under which an open trap can be used to realize thermonuclear regimes with $Q \approx 1$ and to formulate directions for the further development of the concept of an axially symmetric open trap without significantly complicating its magnetic system. Regimes are considered in which the injection of high-energy (fast) neutral particles into the plasma is the only or at least the main source of plasma heating. Such a powerful injection of fast particles allows the creation of positive potential barriers at the ends of the system that contribute to a marked improvement in longitudinal confinement.

Two modes of longitudinal losses of ions can be distinguished: collisional and kinetic (collisionless) [95–97]. In the first case, the trap is filled with a 'warm' Maxwellian plasma, and the rate of plasma outflow through the 'end mirrors' is approximately equal to the speed of sound, i.e., the outflow is gasdynamic. Such a system is called the gas-dynamic trap (GDT) [98, 99]. With a sufficiently large length of such a trap filled with deuterium–tritium (D–T) plasma, the power of a thermonuclear reaction can exceed the energy losses associated with the escape of plasma through the ends. The required length of the reactor is ~1000 m [98, 99], which, apparently, severely limits the energy use of such systems. A promising project is a compact neutron source based on GDT [100].

In the experiments at the GDT installation at the G.I. Budker Institute of Nuclear Physics, in addition to the gas-dynamical regime of plasma outflow, the ambipolar confinement regime in the scheme with a compact pluggin CMS was also demonstrated [101]. In this case, it is possible to suppress the losses by more than an order of magnitude, which substantially reduces the system size requirements.

The injection of fast neutral atoms also contributes to the formation of a positive potential barrier at the ends of the system. The ions that form as a result of ionization experience rare collisions in comparison with the 'warm' ions. The angular scattering of fast ions is much less intense than the scattering of ions from a 'warm' plasma. At the ends of the system, near the points of reflection of fast particles from magnetic mirrors, regions of increased plasma density and a positive potential barrier are formed. Therefore, in the presence of a significant population of fast ions, the regularities of the outflow of such a plasma from the trap are very different from the gasdynamic regime. The larger the population of fast particles, the greater the height of the potential barrier. In order to ensure a high density of fast particles, in addition to the high injection power, it is also necessary to maintain a sufficiently high electron temperature T_e in the trap. Since the slow-down time of fast particles is proportional to $T_e^{3/2}$, to maintain the population of fast particles at the level of the Maxwellian population it is necessary to raise T_e up to 10–20 keV with respect to the $T_e \sim 1$ keV characteristic for the gas-dynamic regime. At high temperatures, the collisionless kinetic regime is realized, and the longitudinal losses become comparable with the longitudinal losses in the ambipolar trap [102].

It should be noted that during the transition from the gasdynamic to the collisionless kinetic regime, the kinetic instabilities characteristic of systems with a loss cone can have a significant effect on the confinement [103]. Stabilization of such instabilities is possible due to the choice of the sizes and geometry of the magnetic field of the trap, the blowing of neutral gas at the ends, and a number of other methods. The global modes of the flute instability are effectively stabilized by the differential rotation of the plasma. In the GDT experiments, the differential $\mathbf{E} \times \mathbf{B}$ rotation was maintained due to the radial distribution of the potential given by the end electrodes. The theory of such a regime, called vortex confinement [104], is confirmed in experiments. Rotation of the plasma as such causes an instability of the 'negative mass' type, but in the case of a variable in the radial angular velocity of the flute instability mode, they are localized near the plasma surface [105]. Differential rotation not only leads to the stabilization of the flute modes, but also to a decrease in the growth rates or suppression of the instabilities of the drift type responsible for the transverse transport. The suppression of dissipative drift instability and transverse transport was demonstrated on the ambipolar GAMMA-10 trap [106, 107]. Even earlier, the stabilizing effect of non-uniform rotation was observed on open traps with the injection of Ogra-I [108] and Alice [109]. Differential rotation can be created without contact of the plasma with the ends, for example, with tangential (with respect to rotation) injection of fast particles.

An important advantage of open traps in comparison with already classical tokamaks is the possibility of stable confinement of plasma with a high ratio β of plasma pressure to magnetic pressure. In particular, in the GDT, the regimes with $\beta \approx 0.5–0.6$ were realized [100, 101].

Simulation of the kinetics of fast ions with reference to the conditions of experiments on GDT was carried out earlier in the approximation of practically complete absence of angular scattering and corresponding losses [110]. In this work, numerical modelling of the ionic kinetics is considered on the basis of the physical model [25, 111], which takes into account the angular scattering of fast particles, as well as their participation in the thermonuclear reactions.

The model of the physical kinetics of fast particles, which is the basis of this study, is described in detail in [25]. The corresponding technique of numerical simulation is presented in [112]. Here we consider in the concise form the formulation of a non-stationary space-zero-dimensional problem with a two-dimensional velocity space in the Fokker–Planck kinetic equation [113].

In the general case, the Fokker–Planck equation is integro-differential [114, 115]. Its solution requires considerable computing resources. In the framework of the approximations adopted in [25, 111, 112], the Fokker–Planck collision operator is represented in a differential form, which significantly reduces the computational

complexity of the problem while maintaining the physical adequacy
of the description of the processes. In this problem, the main role
is played by high-energy (fast) ions, for which collisions can be
taken into account only with particles of thermalized populations.
When calculating such collisions, the thermalized populations can
be approximately assumed to be Maxwellian. Fast particles after
slow-down also form a thermalized population, for which an iterative
process is organized in time with a small number of iterations. In
the case under consideration, there are at least two sources of fast
ions: injection of neutral beams and a thermonuclear reaction. The
products of the D–T reaction are α-particles produced with an energy
of 3.52 MeV. Therefore, the collision operator takes into account
both Coulomb collisions and elastic nuclear collisions at energies
exceeding 1 MeV [116].

In the coordinates, the velocity and the angle between the velocity
vectors of the particle and the magnetic induction, the Fokker–Planck
equation for the distribution function f_a of particles of type a is
written in the form

$$\frac{\partial f_a}{\partial t} - \frac{1}{v^2}\frac{\partial}{\partial v}v^2\left[D_{vv}^C\frac{\partial f_a}{\partial v} - (A_v^C + A_v^N)f_a\right] -$$

$$-\frac{1}{v^2\sin\theta}\frac{\partial}{\partial\theta}\sin\theta D_{\theta\theta}^C\frac{\partial f_a}{\partial\theta} = \frac{s_a(\theta)}{4\pi v_{0a}^2}\delta(v - v_{0a}) - L_a, \qquad (2.7)$$

where D_{vv}^C, $D_{\theta\theta}^C$, A_v^C are the Coulomb diffusion coefficients and
Coulomb dynamic friction; A_v^N is the coefficient of friction due
to elastic nuclear scattering; $s_a(\theta)$ is the angular distribution of the
source; L_a is the particle loss operator.

In the stationary case, $\partial f_a/\partial t = 0$, but, for example, to take
into account the thermalized population of injected particles,
the time evolution of the solution of the non-stationary Fokker–
Planck equation is considered. The dependence of the source
of particles on velocity is adopted in the form of a δ-function,
the angular distribution of the source is normalized as follows:

$\frac{1}{2}\int_0^\pi s_a(\theta)\sin\theta d\theta = q_a$, where q_a is the number of particles formed in

the trap volume per unit time. The particle losses are represented
by the transverse and longitudinal components with respect to the
magnetic field $L_{a\perp} = f_a/\tau_\perp(v,\theta)$ and $L_{a\parallel} = f_a/\tau_\parallel(v,\theta)$. The transverse

loss time $\tau_{\perp}$ is determined by the turbulent transport of particles across the magnetic field; in the calculations it is assumed constant. The longitudinal loss time $\tau_{\parallel}$ of the particles trapped in the loss region in the velocity space is equal to the transit time along the trap; outside the loss region $L_{a\parallel} = 0$. For distribution functions of background particles close to Maxwellian, one can use the diffusion and friction coefficients corresponding to the Rosenbluth–Trubnikov potentials for the Maxwellian background [113, 117].

Knowing the cross sections for elastic nuclear scattering, for example, from experimental data [116], the coefficient of dynamic friction A_v^N can be represented in the form $A_v^N = -\frac{1}{2}v\sum_b nb(\sigma v)_b(\delta E/E)_b$, where $(\sigma v)_b$ is the product of the scattering cross section for the velocity of a 'fast' particle, $(E/E)_b$ is the fraction of energy transferred in scattering of a particle of grade b. Assuming scattering isotropic in the angles, the particle losses associated with this process can be represented as follows: $L_a^N = (1-\cos\theta_L)f_a\sum_b n_b(\sigma v)_b,$, where L is the angular size of the loss region at the given velocity.

In the presence of a symmetric loss region, the distribution function is subject to the following boundary conditions and the symmetric conditions: $f_a(v > v_{0a}, \theta) = 0$, $f_a(v, \theta) \approx 0$ in the field of losses, $\frac{\partial f_a}{\partial v}(v=0,\theta) = 0$, $\frac{\partial f_a}{\partial \theta}(v,\theta=0) = 0$, $\frac{\partial f_a}{\partial \theta}(v,\theta=\pi) = 0$.

The region of losses in the phase velocity space is given by the condition

$$\frac{m_a v_{\parallel}^2}{2} > \frac{m_a v_{\perp}^2}{2}\left(\frac{B_m}{B_c}-1\right)+Z_a e\Delta\phi, \tag{2.8}$$

where e is the electron charge; m_a and Z_a are the mass and charge of the particles of the considered variety a; $\Delta\varphi$ is the electrostatic barrier for particles of type a; B_c is the magnetic field induction in the central part; B_m is the magnetic field induction in 'magnetic mirror cells; $v_{\parallel} = v\cos\theta$ and $v_{\perp} = v\sin\theta$ are the longitudinal and transverse components of the particle velocity.

Taking into account the diamagnetism of the plasma, we have $B_c = B_0\sqrt{1-\beta}$, where B_0 is the vacuum value of the induction of the magnetic field of the central solenoid. The plasma pressure in the 'magnetic mirrors' is considered low, therefore the value of B_m corresponds to the magnetic field of the 'magnetic mirrors' simultaneously in vacuum and in the plasma.

For approximate estimates and comparison with numerical results, one can use a simple analytical solution for the distribution function in the epithermal energy range obtained in [14],

$$f_{0a}(v) = \frac{q_a \tau_{sa}}{4\pi (v^3) v_{ca}^3},$$ (2.9)

where $\tau_{sa} = 6\pi\sqrt{2\pi} \dfrac{m_a}{\sqrt{m_e}} \dfrac{\varepsilon_0^2 (k_B T_e)^{3/2}}{\Lambda_{a/e} Z_a^2 e^4 n_e}$ is the slow-down time due to Coulomb collisions; $v_{ca} = \left(\dfrac{3\sqrt{\pi}}{4} \dfrac{\Lambda_{a/i}}{\Lambda_{a/e}} \dfrac{m_e}{n_e} \sum_i \dfrac{Z_i^2 n_i}{m_i}\right)^{1/3} \left(\dfrac{2K_B T_e}{m_e}\right)^{1/2}$ is the critical velocity at which the slow-down on the electrons is equal to the slow-down on the ions, the summation is made over all types of ions i.

Expression (2.9) corresponds to a number of simplifications. It is applicable in the velocity range $v_{T_i} < v < v_{T_e}$ where v_{T_i} and v_{T_e} are the thermal velocities of the Maxwellian ions and electrons. Any losses, including those associated with angular scattering, are not taken into account. This suggests that the fast particles completely transfer energy to the Maxwellian components upon slow-down to $v \sim v_{T_i}$.

As a criterion of optimization for initial estimates, the ratio of the power P_{inj} injected into the plasma with high-energy particles to the fusion power P_{fus} was considered. For an effective neutron source, it need not be much greater than unity. An approximate analysis using (2.9) yields the following optimal values: $T_i \approx T_e \approx 10$ keV (T_i is the ion temperature), the initial energy of the injected particles is about 100 keV. We note that these results differ significantly from the results obtained by numerical simulation, which indicates a significant effect of the losses associated with angular scattering and the need for their correct recording.

The code FPC2 was used [112] for the numerical modelling of plasma kinetics in modes with injection of high-energy particles, the main module of which is intended for the numerical solution of the Fokker–Planck equation (2.7). The code provides a procedure for calculating the concentration of fast particles along the axis of the trap, which makes it possible to determine the increase in density near the 'magnetic mirrors' and the corresponding ion electrostatic barrier $\Delta\varphi$. At this stage, this procedure was not used, since the value of $\Delta\varphi$ depends on the specific design of the end sections of the magnetic system and the spatial distribution of the injection. The values were set taking into account the experimental data. For example, the distribution of the yield of the D–D reaction measured in GDT [118] shows that the plasma density near the 'magnetic mirrors' can be about 3–5 times the density in the central section of the trap. This corresponds to $e\Delta\varphi = (1–1.5)kT_e$, where k is the Boltzmann constant. To increase this value it is possible in the constructive scheme with injection into the end plugs. At $e\Delta\varphi \leq kT_i$, a noticeable stationary population of 'fast' particles is formed which, after slow-down, form a stationary Maxwellian population. With a further increase of the potential barrier, the Maxwellian component begins to dominate. An important effect associated with the presence of a significant amount of fast particles is the increase in the reaction rate in comparison with a Maxwellian plasma, for example, at $T_i \approx T_e \approx 20$ keV and $e\Delta\varphi = 0.5kT_e$, the reaction rate is twice as high as the velocity in the Maxwellian plasma, at $e\Delta\varphi = kT_e$ – it is 1.5 times higher, the difference is 10% at $e\Delta\varphi = 2kT_e$. At lower temperatures the effect is stronger.

Here we considered regimes with injection of only one kind of particles – either deuterium or tritium. The second component of fuel was considered Maxwellian. The calculations showed practically the same efficiency for a given energy both for the case of deuterium injection only and for the case of tritium injection only. The results given below correspond to the second case. An example of the distribution function in coordinates is the velocity v and the angle θ obtained as a result of the numerical solution shown in Fig. 2.3.

The stationary distribution is practically established when the time counted from the start of injection reaches the value $t \approx 3\tau_s$, where the value of τ_s is calculated from the parameters of the stationary regime. According to numerical results, the optimal value of the energy of the injected neutral particles in the considered regimes is about 250 keV, which is much higher than the approximate estimates.

The energy balance is shown in Fig. 2.4. The diagram shows the following components of the energy balance: P_{fus} is the power of fusion energy release; P_n is the power in neutrons (80% of P_{fus}); P_{ext} is the power of external heating; P_{br} and P_s are the power of bremsstrahlung and sychrotron radiation; P_{ch} is the loss power with charged particles; $(P_{fus})_i$ is the power transferred from fusion products (α-particles) to ions; $(P_{fus})_e$ is the power transferred from the products to the electrons; P_{ie} is the power exchange energy between

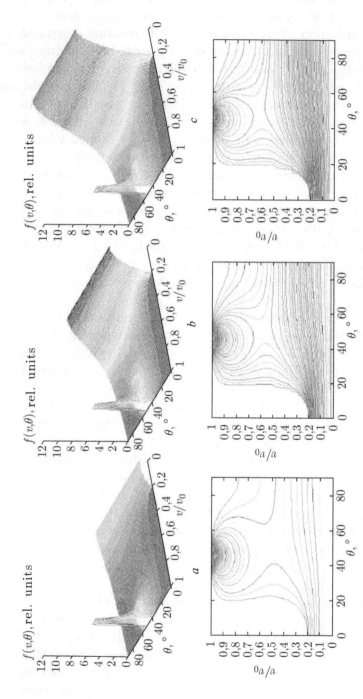

Fig. 2.3. The distribution function of the tritium ions (top) and its contours (bottom) at the instants of time $t = 0.1\tau_s$ (a), $0.3\tau_s$ (b) and $10\tau_s$ (c). The deuterium density $n_D = 3.3 \times 10^{19}$ m^{-3}, the initial energy of the injected particles is 250 keV, the injection angle is $45° \pm 5°$, the injection power is 2 MW/m^3, $T_i = T_e = 20$ keV, the ion electrostatic barrier $\Delta\varphi = 10$ kV, the slow-down time with respect to the parameters of the stationary regime is $\tau_s = 4.5$ s, the transverse loss time $\tau_\perp = \tau_s$

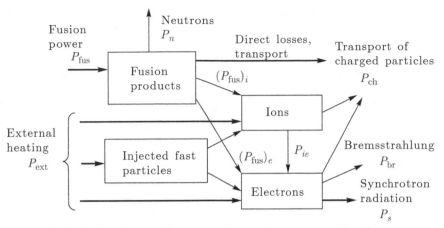

Fig. 2.4. The scheme of energy balance of a thermonuclear plasma.

Maxwellian ions and electrons. According to the kinetic calculations, in the considered regimes the energy losses of the injected particles in the angular scattering can be up to half of the total injected power, and the energy loss of the α-particles (including the production in the loss region) is about 20–30% of their initial kinetic energy. Longitudinal confinement and losses of the Maxwellian components are calculated according to [95–97, 102]. The transverse losses in our model are characterized by a constant time comparable to τ_s; while they are an order of magnitude smaller than the longitudinal losses. Radiation accounts for an insignificant share in the total balance (about 2% of P_{fus}). According to the results of calculations for both Maxwellian particles and for fast particles, the longitudinal losses dominate over the transverse ones, which is a typical situation for open magnetic traps.

Table 15 presents four versions of the results of energy balance calculations, in which the main (and in variants *1, 3, 4* – the only) source of external plasma heating are injected fast particles. In option 2, about a quarter of the external power comes from electron cyclotron resonance (ECR) heating. It can be noted that the highest efficiency is achieved at sufficiently high ion potential barriers $\Delta\varphi$. To create them, one may need to complicate the end parts of the magnetic system. A high neutron yield of N and $Q \approx 1$ can ensure the competitiveness of the presented neutron source on the basis of an open trap among similar systems and other concepts [60, 90, 119, 120]. The fluxes of neutron energy and heat fluxes from the plasma reach high values, which at the same time seem acceptable for a sufficiently long operation of the system under consideration.

Table 15. Parameters of an open trap with a plasma radius $a = 1$ m, length $L = 10$ m and $\beta = 0.5$

Parameter	Options			
	1	2	3	4
The magnetic field of the central solenoid B_0, T	1.5	1.5	2	2
Magnetic field in mirrors B_m, T	11	11	14	14
Deuterium density n_D, 10^{20} m^{-3}	0.22	0.26	0.21	0.415
Density of tritium n_T, 10^{20} m^{-3}	0.33	0.26	0.42	0.415
Density of α-particles n_a, 10^{20} m^{-3}	0.04	0.03	0.06	0.085
Ion temperature T_i, keV	11	10	22	22
Electron temperature T_e, keV	8.5	10.5	18	19
Ion electrostatic barrier $\Delta\varphi$, kV	16.5	15	33	44
The initial energy of the injected particles ε_0, keV	250	250	250	250
Average energy of injected particles $\langle\varepsilon\rangle$, keV	90	90	100	65
Injection power P_{ink}, MW	74	60	60	55
Power of ECR heating P_{RH}, MW	0	18	0	0
Power in neutrons P_n, MW	30	24	43	59
Plasma gain $Q = P_{fus}/(P_{inj} + P_{RH})$	0.5	0.38	0.9	1.34
Neutron yield N, 10^{18} neutrons/s	13	11	19	26.5
The energy flux of neutrons from the plasma J_n, MW/m^2	0.4	0.4	0.7	1
Heat flux from the plasma J_H, MW/m^2	1.8	1.2	1.8	2.0

Field reversed configuration (FRC)

3.1. Introduction to FRC

The field reversed configuration (FRC) combines the properties of both closed and open traps [121]. The plasma in the FRC is almost completely located in the region of closed magnetic field lines bounded by the separatrix (see Figs. 3.1, 3.2). Behind the separatrix is the area of open lines. The magnetic field of FRC is usually considered to be purely poloidal. This means that the magnetic lines lie in the r–z plane, the toroidal component of the magnetic field (along the azimuth θ), as a rule, is absent.

In the literature they often refer to the article by A.S. Kolb and others [122] as the first work on the FRC, but it does not contain a reference to the reversal of the field. In fact, experiments with FRC (as a separate direction) began after R.K. Linford visited the

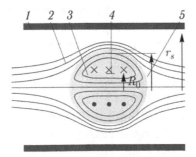

Fig. 3.1. Field reversed configuration (FRC): *1* – vacuum chamber/coil, *2* – open magnetic field lines, *3* – closed field lines, *4* – neutral layer, *5* – separatrix. Crosses and dots show the direction of the azimuth current.

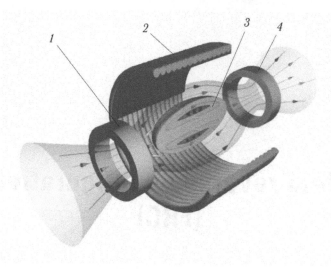

Fig. 3.2. Plasma in the closed magnetic lines of the FRC system with a central solenoid and locking plugs: *1* and *4* – magnetic coils at the ends of the installation, performing the functions of plugs, *2* – solenoid, *3* – plasma supported by azimuth current, inside the separatrix.

USSR and saw the experiment of R.Kh. Kurtmullaev on a pinch with a reversed field [123], after which he launched the FRC research program at the Los Alamos National Laboratory (LANL) [124] and for many years supervised the entire magnetic thermonuclear program in the United States. Therefore, until now the most complete reviews on FRC are the work of two authors: R.Kh. Kurtmullaev [7] and M. Tuszewski [125].

In the Russian literature, the structure with a purely poloidal magnetic field is described by the term 'compact torus' [7] or 'compact toroid' is used, while a system having an essential toroidal field is called a 'combined compact torus' [7] or 'spheromak' [125, 127]. Earlier, the concept of 'compact toroidal torus' was also used [128]. In the foreign literature, compact tori include the spherical tokamak, FRC and the spheromak [129]. Here the authors use the English term FRC [125, 130, 131].

The idea of FRC originated in the 70s of the XX century. Currently, the FRC is being investigated at facilities in Russia, the United States, Japan and other countries. Data on some installations are presented in Table 16.

The simplest spherical FRC model is called Hill's vortex [132]. FRC is sometimes called a compact toroidal configuration [123],

Table 16. General information on existing installations of FRC type

Installation	Location	Main research areas
Boulder/ Colorado FRC	University of Colorado, Boulder	Investigation of plasma turbulence, study of the phenomena of transport of particles and energy in confined plasma
C-2, CBFR Tri Alpha	University of California, Irvine	Experiment on colliding beams, neutron-free $p–^{11}$ B reaction
FIREX	Cornell University, Ithaca	Study of equilibrium problems
FIX NU-CTE-3	University of Osaka, University of Nihon	FRC formation based on the θ-pinch, neutron source
FRX-L, FRCHX Shiva-FRC	LANL, Los Alamos AFRL, Kirtland	Magnetized target fusion (MTF), high density and temperature, solid liner
KT, TOR, TOR-LINER	TRINITI, Troitsk	Investigation of heating, compression, and translation of a compact torus
OMAK	FIAN, RAS, Moscow	Forming the configuration, reversing the magnetic field
Princeton FRC, MRX, SPIRIT	PPPL, Princeton	Investigation of the processes of reconnection of magnetic field lines, stability
ROTAMAK	University of Flinders, Australia	Maintenance of current by a rotating magnetic field (RMF), study of spherical configurations
SSX	Swartmore, Pennsylvania	Forming FRC by merging spheromaks
TCSU, STX, TRAP, PHDX, IPA	Washington University, Seattle, Redmond Plasma Physics Laboratory	Investigation of a rotating magnetic field, acceleration of FRC to sustain tokamak plasma, obtaining a high plasma density
TS-3, 4	Tokyo University	Merge of toroidal configurations
XOCOT	University of Michigan	Application as propulsion system

because it has a toroidal geometry (closed magnetic field lines), but also open field lines of force. In FRC, just like in a tokamak, a huge current flows through the plasma, but it does not create a holding poloidal magnetic field (as in a tokamak). The poloidal field is formed in the FRC by means of external coils, the azimuthal current is a consequence of the reversal of the magnetic field.

The magnetic field in the FRC is defined as

$$B_0(r) = B_e \cdot \sqrt{1 - \beta(r)}, \tag{3.1}$$

where B_0 is the field in the median plane on the axis, B_e is the external magnetic field that is not equal to the field in the coil $B_e \neq B_c$ (the value of the magnetic field at the plasma boundary, where $p = 0$), and β is the ratio of the internal magnetic field to the external magnetic field field equal to

$$\beta(r) = p(r) / (B_e^2 / 2\mu_0), \tag{3.2}$$

where p is the pressure, $\mu_0 = 4\pi(10^{-7}) = 1.26 \cdot 10^{-6}$ H/m is the magnetic constant.

The high value in the purely poloidal FRC on the magnetic axis $\beta \sim 1$ is the limiting value for systems with magnetic confinement. In addition to β, from the equation (3.2), the important parameter is also the Barnes beta [133]

$$\langle \beta_B \rangle = 1 - x_s^2 / 2, \tag{3.3}$$

where x_s is the ratio of the separatrix radius r_s to the radius of the magnetic coil r_c. The allowable plasma pressure is calculated as the product of the external magnetic pressure and β_B.

The high energy content of the plasma, resulting from large β values, makes FRC the first candidate for a low-radioactive thermonuclear reactor on a D–^{3}He mixture.

3.2. Physics of FRC

Modern achievements of experimental and theoretical studies of FRC physics are reflected in the review [134]. From the analysis of the experimental data [7, 125, 134, 135, 144] it follows that typical for FRC experiments are the following values of the parameters: the separatrix radius $a \approx 0.15$ m, the external magnetic field $B_e \approx 0.5$ T, the so-called total temperature $T = T_i + T_e \approx 400$ eV (T_i is the ion temperature, T_e is the electron temperature), the parameter β near the separatrix $\beta_s \approx 0.5$ and more. The confinement times of the energy τ_E and the magnetic flux τ_f are approximately equal to the confinement time of the particles τ_N. For most experiments, $T_e/T_i \approx 0.5-1$, in some cases $T_e/T_i \sim 0.1$. The highest FRC parameters achieved in modern experiments are as follows: $B_e \approx 1$ T, $a \approx 0.2$ m, $T_i \approx T_e \approx 1$ keV [134, 138, 141, 142, 144].

The traditional way to create FRC in an experiment is to reverse the field to a theta pinch. The new C-2 device uses the merging of two counter-toroids formed in theta-pinch chambers [144, 145]. The same method of formation in combination with the acceleration of counter plasmoids is provided in the FRC system with inductive plasma accelerators [146, 147]. In the TS-3 and TS-4 installations, merging of spheromaks is used [148, 149]. These methods imply a multi-chamber (in the simplest case, a two-chamber) installation scheme. The processes of formation, acceleration, compression, confinement, and others are realized in special sections (chambers) of the installation. Multichambers schemes allow one to effectively insert energy into the plasma at various stages of the process and obtain extremely high FRC characteristics at the final stage. Inductive formation in a single-chamber scheme can be realized with a high efficiency of energy transfer of the storage device to plasma formation [150].

Significant progress was made in the development of experimental methods for prolonging the existence and improvement of plasma confinement in FRC [134]. The method of current maintenance by a rotating magnetic field is implemented in the TCS [151–153]. In experiments at the FIX facility, when neutral particles were injected, the plasma confinement time was several times increased [154–156]. Generation of shear flows (forced and spontaneous) was observed in experiments on TS-3 and TS-4 [149].

Let us note one more important moment of experimental physics of the FRC. Usually, the magnetic field of the FRC is purely poloidal and does not have the crossing of the lines of force (magnetic shear). But in a number of experiments, regimes of generation of a small toroidal component of the magnetic field in FRC were observed [157, 158]. The existence of equilibrium configurations of FRC with a weak toroidal field and magnetic shear was also theoretically demonstrated [159, 160].

For most processes and phenomena in the FRC, adequate theoretical models and methods of analysis have been developed [134]. The most important of the elementary properties of the trap is the absolute confinement of charged particles in the magnetic configuration, that is, confinement without taking into account their collisions and participation in collective interactions. The criteria for absolute confinement follow from the analysis of particle trajectories.

The classification of trajectories in FRC and the region of losses in phase space was considered in [161–163]. A criterion for the effective confinement of high-energy thermonuclear products was obtained in Ref. [164]. The stochasticity of high-energy particles (violation of the adiabaticity conditions of motion) and the associated possible losses were analyzed in [165–167].

Summarizing the properties of absolute confinement in the FRC, we can say that the external magnetic field and the radius of the separatrix should be as large as possible. For example, to keep 14 MeV protons practically without any initial losses (the so-called first orbit losses), the product of these quantities must satisfy the condition $B_e a > 15$ T·m, for alpha particles with an energy of 3.5 MeV $B_e a > 5.5$ T·m [164]. The dynamics of high-energy (fast) particles is the determining process in the simulation of injection [168] and maintenance of the current by fast injected particles [169]. Fast thermonuclear products, besides heating the plasma, can also contribute to the maintenance of current [170].

The equilibrium structure of a plasma in FRC [133] is usually simulated using the Grad–Shafranov equation in the ideal magnetohydrodynamics (MHD) approximation. The formulation of the problem in this case requires setting the plasma pressure $p(\psi)$ dependence on the magnetic flux function. Such a dependence is usually given in some form consistent with the experimental data on the structure of the magnetic field and the pressure distribution in the FRC. A multi-fluid equilibrium theory was developed in Ref. [159]. Important theoretical results obtained relatively recently are related to the analysis of a radial static electric field, shear flows and zonal flows in FRC [149, 160, 171, 172]. As is known, an inhomogeneous static electric field and shear $\mathbf{E} \times \mathbf{B}$ flows play an important role in suppressing turbulence and forming regimes of improved confinement, which is a common property of a plasma held by a magnetic field.

MHD modelling of the FRC dynamics is an extremely complex task, requiring the consideration of processes of various nature. The FRC plasma evolution model, taking into account the various loss channels [173, 174], makes it possible to analyze the effect of transport mechanisms on the dynamic FRC structure under given spatial laws describing losses. The study [175] considered the FRC simulation on the basis of a two-fluid MHD code, which allows for the possibility of taking into account the anisotropy of the thermal conductivity and the viscosity of the plasma.

A high level of understanding of the corresponding MHD phenomena has been achieved in the theory of MHD stability of the FRC. It should be noted that in FRC experiments the time of stable plasma confinement is much larger than the characteristic times of development of MHD instabilities. This, apparently, is associated with stabilization factors.

In many experiments, the time of stable confinement is limited by the development of rotational instability, the development of which has been reliably detected in many FRC installations [145, 176–178]. Decrease in the growth rates of rotational modes by an order of magnitude was observed when creating the sheared rotation [149]. Reference [179] shows the stabilizing effect of a rotating magnetic field. In the experiments of R.Kh. Kurtmullaev and his colleagues [7] there were no signs of rotational instability. This seems to be due to the careful optimization of all processes and the suppression of various loss channels at the initial stage of FRC formation.

Convexity of magnetic field lines in the FRC, at first glance, should stimulate the development of flute instability. As shown in [180], the flute modes can be stable in both elongated and compressed configurations. An additional factor contributing to stability is the current density distribution.

MHD stability with respect to permutational modes was theoretically considered in [181]. The stabilization conditions found were much more stringent than the conditions observed in the experiments, which, apparently, is due to kinetic effects. Factors contributing to stabilization are also magnetic mirrors at the ends of the system and a weak toroidal field. The permutational modes can also be stabilized by a rotating magnetic field [182].

The tilt-mode in experiments is stable or develops to the saturation level [183]. Stabilization of the tilt-mode can be achieved with increasing β on the separatrix [184], increasing the elongation of the configuration [185], generating the toroidal component of the magnetic field [186].

We also note that recently non-linear models [186, 187], as well as techniques based on the analysis of minimum energy states, have been developed for the analysis of MHD stability [188].

Anomalous turbulent transport and microinstability generating it are the most serious problems of FRC physics both from the point of view of experimental measurements of turbulence characteristics and from the point of view of theoretical interpretation of the observed transport level. At the moment, there is no unambiguous answer to the question as to which instabilities cause abnormal transport in the FRC. In some works, theories based on drift-dissipative instabilities were considered for the analysis of anomalous transport [189–191]. However, according to [192], this type of instability should not develop in FRC. Quite a large number of theoretical papers were devoted to the analysis of lower hybrid drift (LHD) instabilities in FRC [192, 193–195], since this type of instability was observed in theta pinch discharges close in the properties to FRC. However, the experimental data actually show the absence of such instabilities in the FRC [195–197].

The most detailed data on the fluctuations in the surface layer of FRC plasma are contained in Ref. [195], devoted to experiments on the TRX-2 installation. But the question of the type of oscillations that cause transport is not given an unambiguous answer in the above work. The measured level of fluctuations in the range of LHD frequencies turned out to be two orders of magnitude lower than the value necessary to explain the observed transport. In the TRX-2 experiments, the frequency range was 10–300 MHz under the following conditions: T_e = 100 eV, T_i = 150–400 eV, a = 4–6 cm, B_e = 0.6–1 T, magnetic field induction on the separatrix $B_s \approx 0.6 B_e$, $\beta \approx 0.6$ [195]. Under these conditions, the position of the maximum of the growth rate of the LHD instability is predicted in the measurement range. However, the maximum signal was recorded in the lowest frequency channel 10–40 MHz, that is, at the lower limit of the measurement range. The corresponding ratio of density fluctuations to the unperturbed density is $\delta n_e/n_e \sim 10^{-3}$. The signal in all other channels (i.e. at higher frequencies) is $\delta n_e/n_e < 10^{-4}$. This level is too low to justify the LHD transport, since for this purpose the indicated value should be $\sim 10^{-2}$ [195].

Analysis of the plasma resistance showed that it varies little over the cross section under experimental conditions [196, 197]. This is also an argument against the hypothesis of LHD transport, according to which the resistance near the separatrix should be much higher than in the inner regions of the plasma.

In [198, 199], the electromagnetic LHD instability was considered for FRC conditions. For $k_\parallel$ = 0, β = 0.6–0.7, $T_e = T_i$, $L_n \sim \rho_{T_i}$, the instability parameters are the following [199]: $k_\perp \sim 1/\rho_{T_e}$, $\gamma \sim \omega_{ci}$. Modes with $k_\parallel \neq 0$ are unstable if $k_\parallel L_n \lesssim 4$ [198]. In [200], an electromagnetic ETG instability with $k_\perp > 1/\rho_{T_e}$ for FRC with $L_n \sim \rho_{T_i}$ was considered. A necessary condition for the instability in this work is the presence of a small longitudinal component of the wave vector $k_\parallel \neq 0$. The maximum growth rate of such modes is comparable to the ion cyclotron frequency, which for $L_n \sim \rho_{T_i}$, in turn, is of the order of the diamagnetic drift frequency.

In [201, 202], propagation of shear and torsion Alfvén waves with a frequency $\omega \sim \omega_{ci}$ excited by an external antenna was studied. At the plasma boundary (on the separatrix), the wave frequency $\omega \approx 0.3\omega_{ci}$. Without noticeable attenuation or growth, such waves penetrated deep into the FRC plasma to the surface at which the ion cyclotron frequency was compared with the frequency of the excited wave ($\omega_{ci} = \omega$). The possibility of heating ions in FRC waves with a frequency of $\sim\omega_{ci}$ and possible mechanisms of resonance damping of such waves are discussed in [203].

We note that under the conditions of FRC experiments, the scale of the electron temperature gradient L_{T_e} is comparable with the scale of the density gradient L_n, the scale of the ion temperature gradient in a number of cases, $L_{T_i} \gg L_n$, i.e., $\eta_e = L_n / L_{T_e} \sim 1$, $\eta_i = L_n / L_{T_i} \ll 1$. In accordance with the classification of drift instabilities, such conditions correspond to ETG instability.

It was suggested that, under the FRC conditions, the development of drift instabilities with maximum growth rate in the ETG range is possible [204, 205].

3.3. Modelling the evolution of FRC

Consider a simplified FRC transport model describing the evolution of the configuration with an allowance for particle and energy losses [206]. The difficulties of modelling are related to the peculiarity of the FRC structure. In the plasma layer located inside the separatrix, the transport mechanism, associated with microinstabilities, is dominant. The loss of particles from the region of open field lines located outside the separatrix is determined by the classical collisional and kinetic processes. As a result, self-consistent gradients of particle concentration and temperature are established on the separatrix, determined to a large extent by the balance of the turbulent particle flux from within the separatrix across the magnetic lines of force and the flow along the open field lines outside the separatrix.

The necessary elements of transport modelling are the structure of the magnetic field and the distribution of the plasma parameters in the magnetic configuration. In the simplest case, the internal structure can be given by the model dependences of the plasma parameters for which boundary conditions are satisfied near the separatrix, formulated on the basis of the balance of the particle fluxes through the separatrix and along the open field lines. The model in question establishes a relationship between the value of the diffusion coefficient on the separatrix and the integral confinement time. As a result, it becomes possible to determine the diffusion coefficient with respect to the confinement time measured in the experiments and make a direct comparison of such values of the diffusion coefficient with the predictions of the theory of drift instabilities.

Modelling of equilibrium and transport in reversed magnetic configurations implies the simultaneous solution of the equations of transport of particles, momentum and energy. In the conditions under consideration, the rate of relaxation of the configuration to hydrodynamic equilibrium exceeds the rate of change of parameters as a result of transport. Therefore, the equation of motion can be replaced by the equation of equilibrium in the approximation of single-fluid magnetohydrodynamics, that is, the Grad–Shafranov equation.

Another standard for modelling transport in the FRC approximation is to consider the transport equations averaged over the magnetic surface [207, 208].

In view of the above, we can consider the equation for the diffusion of particles in a form typical, for example, of the theta pinch:

$$\frac{\partial n}{\partial t} - \frac{1}{r}\frac{\partial}{\partial r}\left(rD_\perp \frac{\partial n}{\partial r}\right) = s_n - \frac{n}{\tau}. \qquad (3.4)$$

Here s_n is the source of particles of the species under consideration, τ is the time of direct convective losses; this time is practically infinite inside the separatrix and comparatively small outside (in the field of open field lines).

The energy equations for ions and electrons in the framework of the indicated approximations have the form

$$\frac{\partial}{\partial t}\left(\frac{3}{2}n_i k_B T_i\right) + \frac{1}{r}\frac{\partial(rJ_i)}{\partial r} = s_{T_i} - P_{i-e}, \qquad (3.5)$$

$$\frac{\partial}{\partial t}\left(\frac{3}{2}n_e k_B T_e\right) + \frac{1}{r}\frac{\partial(rJ_e)}{\partial r} = s_{T_e} + \sum_i P_{i-e} - P_b - P_s. \qquad (3.6)$$

Here, s_{T_i} and s_{T_e} are the sources of heating and energy loss of the corresponding components with the exception of radiation losses and energy exchange between the components.

The energy fluxes are related to particle flows by approximate relations $J_i \approx \dfrac{3}{2} k_B T_i \left(-D_\perp \dfrac{\partial n_i}{\partial r} \right)$, $J_e \approx \dfrac{3}{2} k_B T_e \left(-D_\perp \dfrac{\partial n_e}{\partial r} \right)$.

We will consider elongated FRCs, for which the transport equations in the quasi-one-dimensional approximation (3.4)–(3.6) are the most correct. With simplified estimates, the energy equation is not considered. It is believed that the relationship between the temperature and concentration gradients is known from experiments. Then one can use the following dependences:

$$p / p_0 = \beta_0, \quad p / p_0 = (n / n_0)^{\eta+1}, \quad T / T_0 = (n / n_0)^{\eta},$$

where p_0, n_0 and T_0 are the values of pressure, density and temperature on the magnetic axis (maximum values corresponding to the region of the zero magnetic field), the value $\eta \approx 1–2$.

Since FRC has cylindrical symmetry, its two-dimensional structure is considered in cylindrical coordinates r, z. In the central section ($z = 0$), the pressure balance equation is satisfied

$$p + \frac{B^2}{2\mu_0} = \frac{B_e^2}{2\mu_0}. \tag{3.7}$$

To determine the distributions of the parameters, one can use model profiles [209] of magnetic induction in the central section, qualitatively corresponding to the experimental regimes. For $r < a$ (a is the radius of the separatrix), these profiles have the following form:

$$B_1 = cB_e u, \tag{3.8}$$

$$B_2 = \frac{1}{2} cB_e (u + u^3), \tag{3.9}$$

$$B_3 = cB_e u^3, \tag{3.10}$$

where $u = 2r^2 / a^2 - 1$, $c = \sqrt{1 - \beta_s}$, $\beta_s = p_s/p_0$, p_s and β_s is the pressure and the parameter β on the separatrix.

For $r > a$, the magnetic induction

$$B = B_e \{1 - (1 - c) \exp[-(r - a) / \delta]\}, \tag{3.11}$$

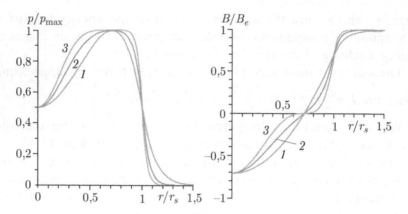

Fig. 3.3. Dimensionless distributions of pressure (*a*) and magnetic induction (*b*) for model profiles *1–3*.

where for the profiles *1–3* given by equations (3.8)–(3.10), the magnitudes of the change in the magnetic induction are, respectively,

$$\frac{\delta}{a}=\frac{1-c}{4c}, \quad \frac{\delta}{a}=\frac{1-c}{8c}, \quad \frac{\delta}{a}=\frac{1-c}{12c}.$$

Profile *1* corresponds to the peak plasma pressure distribution characteristic for the 'rigid rotor' model, profile *3* to the prospective regime with an edge transport barrier, profile *2* to the intermediate mode. Figure 3.3 shows for the profiles *1–3* the distribution of plasma pressure and magnetic induction along the radius in the central section.

To analyze the 2D FRC structure, it is necessary to obtain a two-dimensional distribution of the magnetic field. For this, the Grad–Shafranov equation is solved

$$r\frac{\partial}{\partial r}\left(\frac{1}{r}\frac{\partial \psi}{\partial r}\right)+\frac{\partial^2 \psi}{\partial z^2}=-\mu_0 r^2 \frac{dp}{d\psi}, \tag{3.12}$$

where ψ is the magnetic flux function that determines the magnetic field components $B_r=-\frac{1}{r}\frac{\partial \psi}{\partial z}$, $B_z=\frac{1}{r}\frac{\partial \psi}{\partial r}$, the value of $dp/d\psi$ as a function of ψ is calculated using the model profiles given by the dependences (4.8)–(4.10).

The dependence $B(r)$ for the central cross section makes it possible to find the pressure $p(r)$ and the magnetic flux function $\psi(r)$ in the central section. As a result, we obtain a parametrically given function

$p(\psi)$. This function and its derivative $dp/d\psi$ depend exclusively on ψ, and they are determined not only in the central section, but also in the entire region under consideration. Note that specifying a definite $p(\psi)$ does not imply the uniqueness of the solution $\psi(r, z)$, which is determined to a large extent by the boundary conditions.

The boundary conditions can be formulated as follows. On the axis of the cylinder $\psi = 0$. On the wall $\psi = \psi_w = \text{const}$, i.e., the wall is considered to be perfectly conducting. In practice, this approximation is valid under the conditions when the conductivity of the wall is much greater than the conductivity of the plasma, which corresponds to the experimental conditions. To ensure the certainty of the longitudinal configuration position, magnetic mirrors are provided at the ends. The shape of the wall is specified by the radii in the central part of r_w and in the mirrors r_{wp}. In the magnetic mirrors, the boundary condition $d\psi/dz = 0$ ($B_r = 0$) is given. In the central plane ($z = 0$), the magnetic flux function at the initial instant of time corresponds to any of the above model profiles.

Some magnetostatic problems may require taking into account the inertia of the electrons, which leads to the need for a modification of the equilibrium equation [210]. For our study focused on transport modelling, the approach based on the classical Grad–Shafranov equation is sufficient. This approach is traditional for transport models and particle motion problems in FRC.

In the numerical solution of the system of equations (3.4)–(3.6), (3.12), it is expedient to use the splitting approach to physical processes. At each time step, the changes in the particle density $n(r)$ and the temperatures $T_i(r)$ and $T_e(r)$ are calculated with the magnetic configuration unchanged. Then, for the new $n(r)$, $T_i(r)$, and $T_e(r)$ found, the $B(r)$ and $\psi(r)$ distributions in the central plane are calculated, which are used in the boundary conditions when $\psi(r, z)$ is found in the next step in time. In problems where the energy equation is not considered, the temperature distribution is found using a given parameter η. The model profiles (3.8)–(3.10) can be used as initial conditions,

To analyze the evolution of the magnetic configuration with allowance for transport under the conditions close to the experimental conditions, we were guided by the values of $\eta \approx 2$.

An example of the calculated two-dimensional structure of the FRC magnetic field is shown in Fig. 3.4. The main parameter on which the distribution of the magnetic field depends is β_s. According

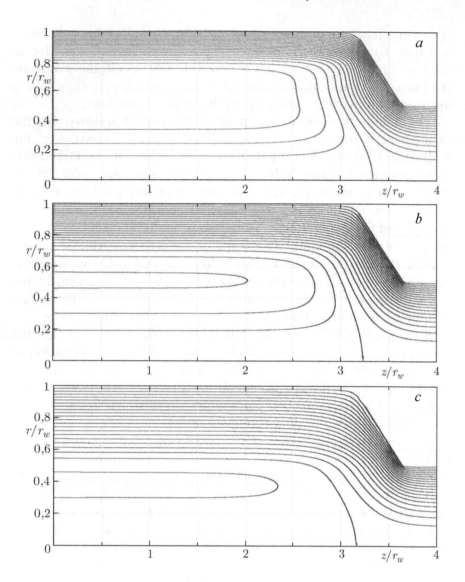

Fig. 3.4. The two-dimensional structure of the FRC magnetic field for profile *3* at $\beta_s = 0.5$ and $a/r_w = 0.8$ (*a*), 0.7 (*b*) and 0.55 (*c*).

to the results of calculations, the shape of the separatrix is determined by the shape of the conducting wall.

It should be noted that often in calculating the structure of the magnetic field in FRC the shape of the separatrix is given, and in the algorithm described above it is in accordance with the boundary conditions on the chamber wall, in the central section and at the ends

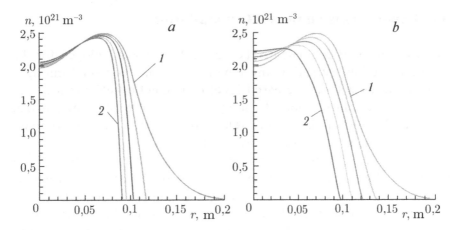

Fig 3 5 Evolution of the particle density in the central section of the FRC at $D_\perp =$ 1 m/s² (*a*) and $D_\perp = 10$ m/s² (*b*): *1* – initial distribution, *2* – at the end of the count at $t = 100$ μs (*a*) and 50 μs (*b*).

of the system. Within the framework of the considered formulation of the problem, the presence of end magnetic mirrors is a necessary condition for fixing a certain longitudinal configuration position. The length of the configuration is determined by the position of the mirrors.

Calculations showed that the characteristic FRC structure with closed magnetic lines of force and separatrix bounded by mirrors is formed at $a > r_{wp}$. These features agree with the results of Refs. [211, 212], in which the influence of mirrors on the structure of the MHD equilibrium of the FRC is investigated.

Figure 3.5 shows examples of calculating the evolution of the particle density under the conditions close to modern FRC experiments. These distributions correspond to the following initial conditions: at the zero-field point, the density of particles (ions and electrons) $n_0 = 2.5 \cdot 10^{21}$ m⁻³ and temperature $T_0 = 1$ keV; an external magnetic field $B_e = 1$ T; the initial distribution of magnetic induction corresponds to profile *1*. The diffusion coefficient is assumed constant, the loss time from the region of open field lines is assumed to be equal to the ion–ion collision time. We note that the experimental data [195–197] indirectly indicate a constant diffusion coefficient in the plasma volume, but its magnitude can vary with time, since the plasma parameters change. The solutions obtained correspond to a zero source of particles $s_n = 0$ in equation (3.4), that is, the regime of free decay of the plasma.

3.4. Experiments on the CTC installation

The experimental equipment CTC [213, 214] was developed for the possibility of investigating and developing a method for forming a compact torus with an increased level of magnetic field capture, and also effective transmission of energy from the battery to the plasma. A general view of the experimental installation of the Compact Toroid Challenge (CTC) is shown in Fig. 3.6.

The proposed scheme is based on the principle of the θ-pinch formation, but it has some differences. One of the common problems of CT generation is the low level of the captured magnetic flux and, as a result, the low level of energy transferred from the energy source to the plasma. The process of configuring is divided into several stages (see Fig. 3.7).

Stage *a*. The main capacitor bank is discharged to the main solenoid, where an azimuth current I_1 occurs. At the same time, a poloidal magnetic field B_p is formed in the chamber. At the moment of reaching the maximum of the current I_1, it breaks.

Stage *b*. After breaking current I_1 in the plasma, the azimuth current I_θ is induced, which tends to maintain a decreasing magnetic field due to the phenomenon of mutual induction. Since the external current I_1 in the main solenoid decreases, according to the Lenz rule, the current in the plasma I_θ will be directed in the same direction. The azimuth current I_θ creates around itself a poloidal magnetic field B_p.

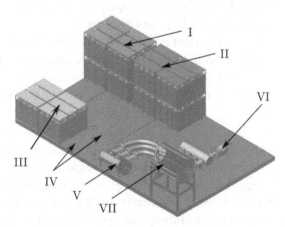

Fig. 3.6. General view of the CTC installation: I – the main capacitor bank (CB1); II – auxiliary battery of capacitors (CB2); III – battery of longitudinal current capacitors (CB3); IV – connecting wires; V – vacuum arresters; VI – vacuum pump; VII – chamber.

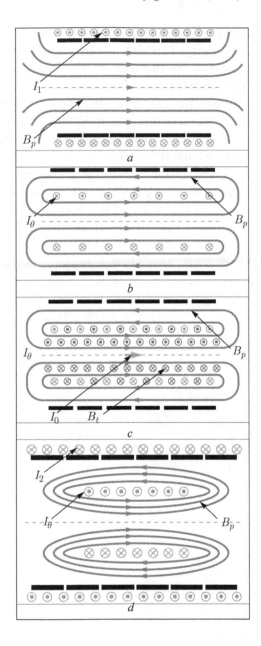

Fig. 3.7. The scheme of formation of a compact configuration: a – current in the winding I_1 leads to the appearance of a poloidal magnetic field B_p; b) after the current I_1 breaks in the plasma, azimuthal current I_θ and poloidal magnetic field B_p appear; c – the current I_0 flowing along the chamber axis creates a toroidal magnetic field B_t; d – squeezing of the plasma from the walls and compression under the action of Ampere's force.

Stage *c*. Simultaneously with the break of current I_1, a longitudinal current battery is discharged, which contributes to the formation of a longitudinal current I_0 on the axis of the chamber. The current I_0 generates around itself a toroidal magnetic field B_t. The total magnetic field becomes helical. The electrons and ions that create the current begin to move along the magnetic lines of force that support the azimuth current I_θ. The resulting configuration, in fact, is a compact torus of elongated shape.

Stage *d*. After a break in the longitudinal current I_0, an auxiliary battery of capacitors is discharged to create a current I_2 in order to squeeze the plasma from the walls of the formation chamber due to the Ampere force between the current I_2 and the azimuth current in the plasma I_θ (see Fig. 3.8).

During the experiments, two plasma regimes were investigated, but the final plasma parameters depended only on the initial level of preionization. For a low level of preionization ($< 5 \times 10^{13}$ cm^{-3}), a rapid penetration of the magnetic field into the plasma occurred

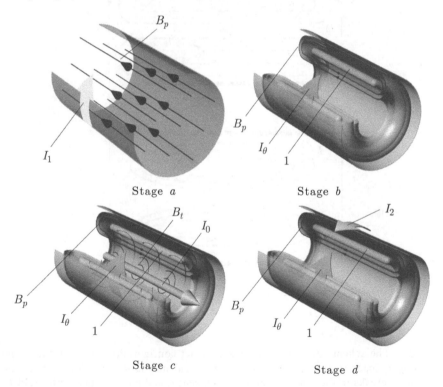

Stage *a*   Stage *b*

Stage *c*   Stage *d*

Fig. 3.8. Three-dimensional scheme for the formation of a magnetic configuration of the compact torus type: *1* – main solenoid, arrows indicate the directions of the currents I_1, I_θ, I_0 and I_2. The stages correspond to the stages in Fig. 3.7.

while simultaneously heating the plasma. As a result, this led to a low value of β for the final configuration (< 0.6). A high level of preionization (> 10^{14} cm^{-3}) increased the magnetic field penetration time by about half, the result was a configuration with a high β. The presented experimental results were obtained in the second mode of operation. In this mode, the measured level of the captured magnetic field B_p was ~0.02–0.03 T. Capacitor batteries were charged up to 25 kV. The peak plasma density was 10^{15} cm^{-3}. The total energy at 25 kV is 150 kJ, the maximum field on the chamber axis at 25 kA is 0.3 T. Hydrogen was used as a working fluid for the production of plasma.

Three independent capacitor banks were used as energy sources. When the first battery was discharged, through a vacuum arrester and a current interrupter, the current flowed through the main solenoid. The second battery is discharged to the auxiliary winding, which provides compression of the configuration in the transverse direction. The third battery causes current flow along the axis of chamber I_0, as well as the creation of a toroidal magnetic field.

The experiments were carried out using two chambers of different diameters. A large chamber is made of a composite material with flanges of plexiglass (chamber 1). The material for the chamber of smaller diameter is quartz, the flanges are made of steel (chamber 2). The quartz chamber was used to obtain a cleaner internal surface, which allows for a deeper vacuum. However, a decrease in the volume of the chamber leads to a decrease in the volume of the plasma, which affects the lifetime of the configuration. A vacuum diffusion vapour-oil pump was used to create the vacuum. The chamber was previously vacuumed for about 24 hours. The flowing poloidal current created a toroidal field in the chamber. The supply cables are made of copper without additional alloying. The detailed characteristics of the experimental setup are presented in Table 17.

Two windings are wound on the chamber with a radius of 0.5 m and a length of 0.85 m – the main winding (49 turns) and the auxiliary winding (48 turns). The windings are connected to the capacitor banks in such a way that the flowing currents were counterdirected. The number of capacitors in the main battery is 22, in the battery of the longitudinal current is 32, in the additional battery is 42. The capacitance of one capacitor is 5 µF. To create the voltage, 10 electrodes located on the flanges were used.

Table 17. Main characteristics of the experimental CTC installation

Parameter	Value	
	Chamber 1	Chamber 2
Number of capacitors in main battery	22	
Number of capacitors in a longitudinal current battery	32	
Number of capacitors in the secondary battery	42	
Capacitor capacitance C, µF	5	
Total energy at 25 kV, kJ	150	
Length, m	0.85	0.8
Radius, m	0.5	0.125
Diameter of wire, m	$5 \cdot 10^{-3}$	
The number of turns in the main solenoid	49	51
Number of turns in the auxiliary solenoid	48	50
Maximum field on the camera axis at 25 kA, T	0.3	

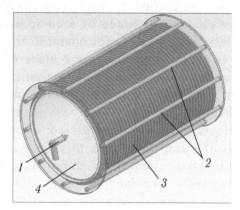

Fig. 3.9. Vacuum chamber with 10 external electrodes: *1* – central electrode, *2* – electrodes around the perimeter of the chamber, *3* – winding, *4* – flange.

Below are photographs of the vacuum chamber (Fig. 3.10) and the experiment (Fig. 3.11) performed by a camera at a shooting speed of 1200 frames per second.

The general scheme of the installation is shown in Fig. 3.12. The final stage of formation of a compact configuration is also presented. The measurements were taken using magnetic probes *5* and Rogowski coils (not shown). Magnetic probes are installed on the axis of the chamber, half the radius of the chamber and near the wall so that

Fig. 3.10. General view of the installation chamber.

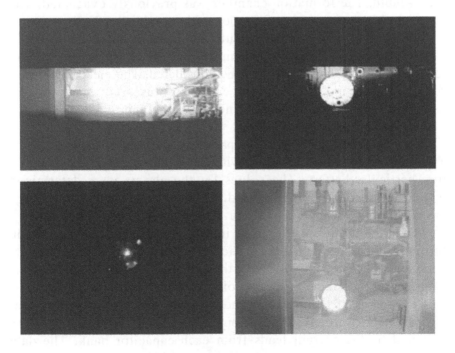

Fig. 3.11. Shooting an experiment.

measurements are made only along the z axis. The Rogowski coils are mounted on the current leads from each capacitor bank. A system for diagnosing plasma parameters based on photo images obtained with the help of a high-speed camera was also tested.

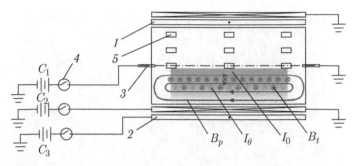

Fig. 3.12. The CTC installation scheme: *1* – main solenoid, *2* – auxiliary solenoid, *3* – central electrode, *4* – dischargers, *5* – probes, C_1, C_2, C_3 – capacitor batteries, B_p – poloidal magnetic field, B_t – azimuthal magnetic field, I_θ – azimuth current, I_0 – axial current.

Hydrogen was used as the working substance for the production of plasma. A vacuum diffusion vapor-oil pump was used to create the vacuum. The formation chamber was previously evacuated. The supply cables were made of copper without additional alloying

The equivalent electrical circuit of the experiment is shown in Fig. 3.13. The circuits for the main and auxiliary solenoids are identical, so only one circuit can be considered. The electrical circuit for passing the poloidal current can be represented as a series of capacitors connected in series and equivalent impedances for the plasma and wires, respectively.

The capacitor bank consists of parallel capacitors of capacitance C. Then the capacitance $C_b = n \cdot C$, where n is the number of capacitors in a particular battery. The values for the plasma were determined when the sensors were calibrated. Typical values of the discharge circuit and plasma parameters are presented in Table 18.

The main purpose of the experiments was to study the formation regime, and therefore the magnetic field was measured and its variation with time and volume of the chamber. For the measurements, magnetic probes and Rogowski coils, mounted on the chamber axis, half the radius of the chamber and near the wall, were used so that measurements were taken only along the *z* axis. Rogowski rings were mounted on the current leads from each capacitor bank. The data obtained from different sensors is presented below.

Figure 3.14 shows the sequence of current pulse transmission through the main solenoid (Fig. 3.14, signal *4*), as well as the longitudinal current through the plasma (Fig. 3.14, signal *3*). The transmission of currents occurs with overlapping in time to ensure the continuity of the process of forming the CT, therefore the separation

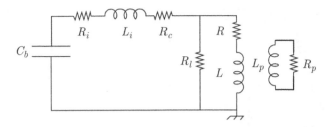

Fig. 3.13. The general electrical circuit for CTC installation: $R_p(t)$ and $L_p(t)$ is the resistance and inductance of the plasma versus time, R_l is the cable load resistance, R_c is the total resistance of all cables, $R_i(t)$ and $L_i(t)$ and the inductance of the arresters on time, R and L are the resistance and capacitance of the winding; C_b is the capacitance of the capacitor bank.

Table 18. Parameters of the discharge circuit and plasma

Parameter	Value
Initial values	
Resistance of plasma R_{p0}, mΩ	4
The plasma inductance L_{p0}, nH	30–100
Winding inductance, nH	4.78
Resistance of cable R_c, mΩ	25
Capacitance of one capacitor, μF	5
Discharge values	
The rate of change in the plasma inductance L_p, nH	7
Rate of change in plasma resistance R_p, mΩ	2
Resistance of the arrester R_p, mΩ	22
Inductance of the spark gap L_p, nH	100–900
After discharge	
Plasma inductance L_p, nG	150–800
Resistance of plasma R_p, mΩ	180–900

at the stage is conditional and made for convenience. As can be seen from Fig. 3.14, unlike the current I_1, the current I_o does not break at the maximum. Its feed stops after a half-cycle of discharge (about 25 μs).

In the course of these experiments, the optimum operating conditions of the plant were determined, the influence of the geometric dimensions of the formation chamber and materials was studied, and also the individual stages of formation were studied. Figure 3.15 shows the results of two series of experiments in which different stages of formation were studied. A series of experiments

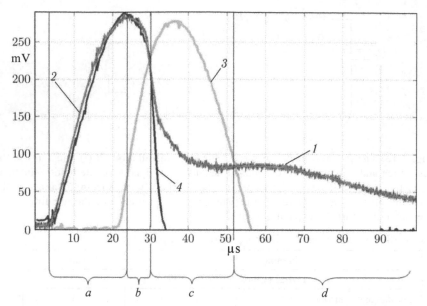

Fig. 3.14. Cyclogram of operation of the installation: *1* – signal from the longitudinal magnetic field sensor in the chamber B_p; *2* – filtered signal from the longitudinal magnetic field sensor in chamber B_p; *3* – signal from the longitudinal current sensor I_0; *4* – signal from the current sensor in the main solenoid I_1. The numbering of the stages corresponds to Fig. 3.7.

consisted of five consecutive discharges in the first case and three in the second. The experiments were carried out with an additional condenser battery and without it, respectively. Figure 3.15 shows the data for each series.

The magnetic field near the wall (Fig. 3.15 *a*, *b*, signal *3*) changes its sign to the opposite, and the magnetic field at the middle of the radius (Fig. 3.15 *a*, *b*, signal *2*) disappears almost to zero. The magnetic field on the chamber axis (Fig. 3.15 *a*, *b*, signal *1*) does not change its sign, which indicates the formation of a reversed magnetic configuration.

Observed in Fig. 3.15 *b* a significant drop in the level of the magnetic field (Fig. 3.15 *b*, signal *1*) is associated with the problem of synchronizing the break of current I_1 and the transmission of the longitudinal current I_0, i.e., in fact, with the difficulty of determining the current break time in the main solenoid due to insufficiently rapid trigger devices. It is important to synchronize the currents, i.e., to coordinate the moments so that the breaking and closing of the currents follow simultaneously. Measurement of currents through solenoids made it possible to control the synchronization of interrupts of different currents. Nevertheless, when the current I_2 passes through

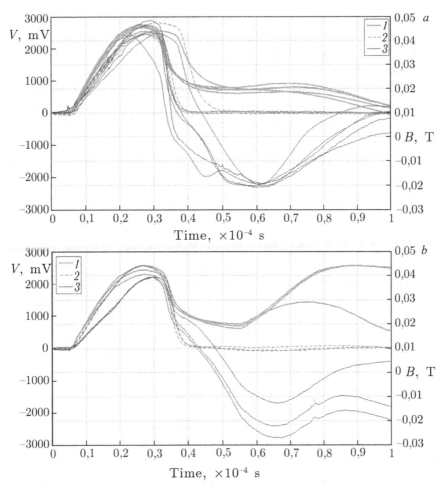

Fig. 3.15. The experimental results of measuring the magnetic field B_p: *a* – a series of five experiments without an additional capacitor bank; *b* – three experimental shots using the discharge of an additional capacitor bank; *1* (solid curves) – change of the magnetic field on the chamber axis, *2* (dotted lines) – magnetic field at the middle of the chamber radius of 12.5 cm, *3* (solid thin curves) – magnetic field on the chamber wall.

an additional solenoid, the magnetic field grows (Fig. 3.15 *b*, signal *1*), caused by an increase in the current in the plasma and a decrease in the configuration size.

Under different experimental schemes, one can obtain an increase in the configuration lifetime or an increase in the level of the captured magnetic flux, and also observe the behaviour of the configuration under these conditions. One of the important effects is the self-heating of the plasma and the increase in the magnetic flux due to pinching of the configuration (the maxima of signals *1* and *3* in Fig. 3.14). The presence of this effect allows one to achieve an increase in the lifetime and the final temperature of the plasma. Such an effect is not observed in a small cell, perhaps it just does not manifest itself enough. The further direction of the experiments should be the unification of various methods for increasing the efficiency of the process, increasing the lifetime and the final temperature of the plasma in one unit.

Figure 3.16 shows the measurement results for the second chamber, smaller diameter. Experiments with a small chamber were carried out without the inclusion of a compression winding. It can be seen that the plasma behaviour differs significantly from the larger camera. After breaking current I_m, there is a gradual decrease in the magnetic flux in the chamber volume. Nevertheless, signal *3* from the wall sensor also changes its sign, which shows us the formation of a reversed magnetic configuration. Signal *1* shows that the magnetic flux retained in the volume after the current break is significantly higher than in the case of the first chamber. This is due both to the improved characteristics of the chamber itself and to the increase in the specific density of magnetic energy in the chamber volume.

Several series of experiments on the formation of the magnetic configuration 'compact torus' with different chambers and plasma parameters were carried out at the CTC facility. The main task was to determine the optimum operating conditions of the installation, the effect of the geometric dimensions of the formation chamber and materials, as well as the study of individual stages of formation. Two series of experiments with discharge and without the discharge of an additional capacitor bank are presented. Since the main purpose of the experiments was to study the formation regime, the main parameter for the measurement was the magnetic field and its variation with time and in the volume of the chamber. The current through the solenoids was also measured, which made it possible to synchronize different currents.

For the experiments, two chambers of different diameters were used. The larger diameter chamber is made of a composite material with plexiglas flanges. The chamber of the smaller diameter is made of quartz with metal flanges. The main reason for using a quartz chamber is the ability to get a cleaner internal surface, and also to achieve a deeper vacuum. However, by reducing the chamber volume and correspondingly the plasma volume, a significant decrease in the configuration lifetime took place.

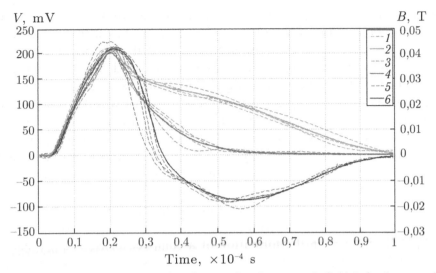

Fig. 3.16. The experimental results of measuring the magnetic field B_t for the second chamber: the change in the magnetic field along the axis of the chamber (*1*), the magnetic field in the middle of the chamber is 6 cm (*3*), the magnetic field on the chamber wall (*5*). Lines *2*, *4*, *6* represent the corresponding averaged values of a series of experiments.

The experiments were carried out with two plasma regimes. The final plasma parameters differed only depending on the level of preionization. With a low level of preionization ($< 5 \cdot 10^{13}$ cm^{-3}), a rapid penetration of the magnetic field into the plasma was observed. Heating of the plasma occurred at the rate of penetration of the magnetic field. However, this led to a low level of β (less than 0.6) for the final configuration. At higher preionization levels ($> 10^{14}$ cm^{-3}), the magnetic field penetrated into the plasma approximately twice as slowly and as a result, a configuration with a high β was obtained.

The use of different experimental schemes allows one to obtain an increase in the lifetime or increase in the level of the captured magnetic flux. Another important effect is the self-heating of the plasma and the increase in the magnetic flux of the configuration due to pinching of the configuration. A similar effect is absent in a small chamber. The further direction of the experiments should be to combine these advantages in one chamber.

Modern high-speed cameras can be used to diagnose various fast processes. This method has several advantages, the most important of which is the possibility of obtaining a two-dimensional picture of the process in real time. A series of similar images allows to investigate the non-uniformity of the plasma. On the other hand, the processing of such images can provide information about the plasma parameters. The plasma parameters diagnostics system based on photo images obtained with the help of a high-speed camera was tested in our experiment by intensity calibration in accordance with known plasma parameters, which were obtained by another method (for example, spectroscopy). In the course of the experiment, the photographing was carried out synchronously with the current transmission, which allowed increasing the resolution for obtaining a more detailed picture of the experiment.

Figure 3.17 shows the distribution of the plasma temperature and the ring structure of the current layer, which confirms the data of magnetic sensors on the formation of a compact torus. The presented result is obtained by applying a two-dimensional median filter to the original image to remove noise, and then a Gaussian filter is smoothed. The temperature is obtained by converting the intensity from the original image to previously calibrated intensity values. To get rid of the effect of preliminary ionization, images without a magnetic field.

3.5. Thermonuclear systems based on FRC

The FRC, like the spherical tokamak, belongs to the class of compact toroidal systems. Therefore, for high β, the parameters of the plasma and the magnetic trap are close in magnitude to the reactors based on these systems. Possible parameters of the FRC reactor on the D–^{3}He fuel were considered in [215–219].

The analysis of the efficiency of the reactor is based on the integral balance equation. The principal difference between the calculation procedure for FRC and the technique for spherical tokamak is the description of the structure of FRC and transport.

Let us determine the sizes of the reactor and the characteristic value of the magnetic induction. From the point of view of transport, the magnetic induction and plasma dimensions should be as large as possible. In addition, as stated earlier, the product $B_s a$ (B_s is the induction on the separatrix, a is the radius of the separatrix) must be at least a certain value corresponding to the effective confinement of fast products of thermonuclear reactions with a minimum initial loss (so-called first orbit losses) [164]. To confine protons with an energy of 14 MeV, practically without initial losses, the product

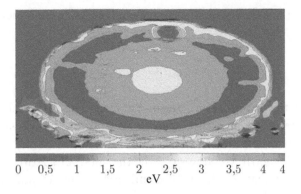

| 0 | 0,5 | 1 | 1,5 | 2 | 2,5 | 3 | 3,5 | 4 | 4 |

eV

Fig. 3.17. Spatial distribution of the temperature of the plasma configuration along the volume of the chamber.

of these quantities must satisfy the condition $B_s a > 15$ Tm·m; for alpha particles with an energy of 3.5 MeV this value is $B_s a > 5.5$ T·m [164].

The greater the induction of the magnetic field, the higher the power of the reactor and the higher energy fluxes from the plasma. The density of the heat flux also increases in proportion to the radius of the plasma. Therefore, the possibilities of the first wall limit the values of B_s and a from above.

The most critical problem in modeling FRC is the theory of collisionless transverse transport. The empirical scaling results obtained can not be considered universal, at least, there are not enough reliable grounds for extrapolation to the area of thermonuclear regimes. Estimating the confinement time of particles requires knowledge of the coefficient of collisionless transverse diffusion. Based on the numerical results obtained within the framework of the theory of electromagnetic gradient drift instabilities, we take for the diffusion coefficient the following theoretical estimate:

$$D_{\perp \text{theor}} = 0.1 \frac{\rho_{T_i}}{L_n} \frac{k_B T_i}{e B_e}, \tag{3.13}$$

where the values of ρ_{T_i}, L_n and T_i correspond to the plasma layer adjacent to the separatrix from the inside.

As a result of the energy balance analysis, the required confinement time can be determined. The developed transport model allows determining the required diffusion coefficient $D_{\perp \text{req}}$ with respect to the confinement time. The confinement enhancement factor is defined as the ratio

$$H_D = D_{\perp\text{theor}} / D_{\perp\text{req}}. \qquad (3.14)$$

This value is the main optimization criterion. The diffusion coefficient can be reduced by generating a shear rotation of the plasma near the separatrix. In this case, the enhancement factor is $H_D \approx 1+(\gamma_s/\gamma)^2$, where γ_s is the shift parameter, γ is the characteristic value of the instability increment. Under FRC conditions, it is possible to maintain a pressure profile providing the values of the shear parameter to $\gamma_s \approx 3\gamma$. Therefore, an acceptable requirement to improve confinement can be considered a decrease in the diffusion coefficient in $H_D = 10$ times compared to the regime without shear flows. This corresponds to an increase in the confinement time by 3–4 times.

For preliminary analysis, we take $B_s = 5$ T, $a = 2$ m for the D–^{3}He reactor and $B_s = 2$ T, $a = 1$ m for the D–T reactor. These values correspond to the products of $B_s a$, which are lower than the values necessary for practically 100% confinement of thermonuclear products, but they ensure the confinement of the vast majority of products. We take the initial losses of fast products equal to 5% of their total number.

As a method of external plasma heating, we consider the injection of fast neutral particles. The energy of the particles is about 200 keV. In this case, the injected particles transmit energy primarily to the ions. The energy released as a result of thermonuclear reactions is transmitted mainly to electrons, since the energies of thermonuclear products exceed several megaelectronvolts (the initial energy of D–^{3}He protons is 14 MeV), and at such energies the slowing down process on electrons dominates. We note that due to radiation, the energy loss through the electron channel is much higher than the losses along the ion channel. The power of heating from thermonuclear and injected particles in most parts is also fed to the electrons. Under such conditions, calculations of the energy balance showed that the temperature difference $\Delta T = T_i - T_e \sim 1$ keV. Since the power P_{ie} of energy exchange between ions and electrons is calculated as a difference of close values, the accuracy of its calculation is not high. The temperature difference ΔT, in turn, is determined by the value of P_{ie}. The estimated absolute error in calculating ΔT is also ~1 keV. Therefore, the temperatures of the ions and electrons can be taken equal to $T_e = T_i$. The wall reflection coefficient of synchrotron radiation wall $R_w = 0.8$. It is assumed that the coating of the first wall is liquid lithium.

Let us consider the regimes of a thermonuclear reactor with a plasma gain $Q = 20$. The results of calculations are presented in Table 20.

Variants FRC-1 and FRC-2 are reactors on D-^{3}He fuel. To show the advantages of FRC as a system for a low-radioactive D-^{3}He reactor, calculations of D–T fuel systems were carried out. The FRC-3 variant corresponds to the parameters of the D–T reactor, the FRC-4 variant – to the parameters of the system with dimensions and magnetic fields close to the current FRC settings.

Table 19 shows the following values: the radius of the separatrix a; length L; elongation of the plasma k; volume of plasma V; magnetic induction on the inner surface of the wall B_e; the beta value on the separatrix β_s and the mean beta $\langle \beta \rangle$; average electron density $\langle n_e \rangle$; the plasma temperature on the magnetic axis T_0 and its average value $\langle T \rangle$; the relative density of plasma components $x_{3\text{He}}$, x_T, x_p, x_α, x_{Li} in units of deuterium density; thermal energy of the plasma E_{th}; fusion power P_{fus}; power in neutrons P_n; bremsstrahlung power P_b; power of cyclotron losses P_s; total radiation losses P_r; the plasma gain Q; the confinement times of α-particles τ_α and protons τ_p; energy confinement time τ_E; H_D confinement factor; the average energy flux of neutrons from the plasma J_n; the average heat flux from the plasma J_H.

In a D-^{3}He reactor based on FRC, the neutron energy flux from the plasma is relatively low (<0.3 MW/m^2), so it can be ensured sufficiently long operation without replacement. The heat flow (about 3 MW/m^2) also seems acceptable.

In the D–T reactor, energy flows to the first wall are relatively high. In addition, in the case of using D–T fuel, as in the case of D-^{3}He fuel, a significant improvement in containment is required. Therefore, the energy advantages of the D–T reaction do not appear under the FRC.

In the conceptual design of the FRC reactor on D-^{3}He fuel ARTEMIS [42], a possible constructive scheme, including energy conversion systems, was also considered, and the cost of electricity generated was calculated, and the competitiveness of the power plant was demonstrated. From a technical point of view, FRC looks almost an ideal candidate for the D-^{3}He reactor, but, unfortunately, to date, the physical processes in the plasma of the reversed magnetic configuration have been studied much less in detail than in the

Table 19. Parameters of thermonuclear systems based on FRC with D–³He fuel (variants FRC-1, FRC-2) and D–T fuel (FRC-3, FRC-4)

Parameter	D–³He FRC-1	D–³He FRC-2	D–T FRC-3	D–T FRC-4
a, m	2.0	2.5	1.5	0.5
L, m	20	20	15	2.5
$k = L/a$	10	8	10	5
V, m³	240	375	101	1.9
B_e, T	5.0	5.0	2.0	1.0
β_s	0.80	0.50	0.50	0.8
$\langle\beta\rangle$	0.93	0.83	0.83	0.93
$\langle n_e\rangle$, 10^{20} m⁻³	5.0	4.6	3.4	1.2
$T_{i0}/\langle T_i\rangle$, keV/keV	67/64	67/59	12/10.6	10/9.5
$x_{^3He}$	1	1	—	—
x_T	0.0064	0.0059	1	1
x_p	0.16	0.13	—	—
x_a	0.34	0.28	0.072	0.0058
x_{Li}	0.05	0.05	0.05	0.05
E_{th}, MJ	3140	4380	190	1.0
P_{fus}, MW	1214	1670	1070	1.57
P_n, MW (P_n/P_{fus})	65 (0.054)	92 (0.055)	860 (0.80)	1.26 (0.80)
P_b, MW (P_b/P_{fus})	628 (0.52)	859 (0.51)	32 (0.04)	0.05 (0.03)
P_s, MW (P_s/P_{fus})	22 (0.017)	67 (0.040)	≈ 0	≈ 0
P_r, MW (P_r/P_{fus})	670 (0.54)	926 (0.55)	32 (0.04)	0.05 (0.03)
Q	20	20	20	0.1
τ_a, s	20	20	3	0.3
τ_p, s	10	10	—	—
τ_E, s	6.3	6.7	0.84	0.06
H_D	2.8	10	10	1.6
J_n, MW/m²	0.26	0.29	6.1	0.16
J_H, MW/m²	2.8	3.2	6.3	0.17

tokamaks. One of the serious questions is related to the anomalous transport in the FRC and the causes that cause it. In particular, in the ARTEMIS project the reserve for anomalous transport is assumed to be such that the confinement time is approximately 200 times less than the classical value. The neutron flux in ARTEMIS is about 0.3 MW/m^2.

Based on the installation of FRC with the sizes and magnetic fields of the level of today's experiments, it is possible to create a thermonuclear system with $Q = 0.1$ operating on D–T fuel. Note that the plasma temperature for this must reach 10 keV. To do this, a complex of heating systems is required, which introduces into the plasma a total power of about 16 MW. This is not too big for today's thermonuclear installations.

The installation with $Q = 0.1$ can be used as a compact source of thermonuclear neutrons. The magnitude of the neutron energy flux of 0.16 MW/m^2 is acceptable for these purposes.

Thus, the practical use of a magnetic trap based on FRC as a thermonuclear system (neutron source) looks realistic for modernized installations of today's level. The most essential condition for the implementation of such systems is an increase in the heating power to tens of megawatts with an energy content of plasma of 1 MJ. The possibility of a stationary or quasi-stationary operating mode for such devices depends on the duration of the operating pulse of the heating system. According to our estimates, in such a system, $\tau_E \sim 0.1$ s, which can be regarded as the minimum required pulse duration.

As a reactor with D–^{3}He fuel FRC, in our opinion, has significant advantages. To fully physically substantiate the concept of such a reactor, it is certainly necessary to experimentally test the predictions of transport simulation in conditions close to a thermonuclear reactor. This requires experiments on new generation FRC units or modernization of existing devices. As a result, the scaling of confinement can be refined, and also the possibility of maintaining the modes of improved confinement is demonstrated. In our model, the cause of transport is considered to be the development of drift instabilities, ie, the standard mechanism for magnetic traps is anomalous transport. Calculations show that, apparently, an improvement in confinement is required in comparison with the predictions of the theory. For this, the methods of generating shear flows can be used, which have been repeatedly demonstrated when creating transport barriers in various magnetic traps. In the FRC, there are all the conditions for generating such flows.

In addition, there is also the possibility of increasing the confinement time without reducing the coefficient of transverse diffusion. In accordance with the model of transport developed for the FRC (Section 3.3), the confinement of the plasma inside the separatrix depends not only on the transverse transport time, but also on the time of longitudinal confinement. We take the longitudinal potential difference in the trap (ion electrostatic barrier) $\Delta\varphi = k_B T_i/e$. In this case, the longitudinal confinement time of the plasma in the region of the open field lines $\tau_\parallel$ is approximately 7 times higher than the ion–ion collision time τ_{ii}. Such a relatively small electrostatic barrier can be maintained by injecting fast particles. The injection of beams of fast particles can be organized in such a way that the orbits of a certain group of particles intersect the separatrix.

Thus, at present, there are no fundamental limitations for the implementation of the D–^{3}He reactor based on FRC. In contrast, FRC has many potential advantages, which, in our opinion, makes this magnetic configuration one of the most preferred plasma confinement systems for the D–^{3}He reactor.

Dipole and multipole traps

4.1. Features of dipole and multipole configurations

Dipole and multipole configurations belong to the class of systems with internal conductors. In such traps there are conductors with current completely surrounded by plasma. The magnetic field in these systems has a poloidal topology. An attractive feature of systems of this class is the ability to provide MHD stability with respect to modes associated with plasma diamagnetism [220]. The simplest example is the dipole configuration created by a single ring current, the circuit of which is shown in Fig. 4.1 *a*. In actual installations, a levitating dipole scheme is used to create a dipole configuration, in which the main coil is held in the horizontal plane by the Ampere force balancing the force of gravity. For this, a second coil with a current of the same direction as in the main coil is used. The largest installations of this type are LDX [221, 220] and RT-1 [223, 224].

Figure 4.1 shows a simple dipole configuration, as well as some advanced configurations. The main direction of optimization of the dipole trap is the creation of a compact configuration. This can be achieved in a two-dipole system [225] or by using several additional coils [226].

A more radical way to control the shape of the plasma is the transition to a multipole configuration with several internal coils. An example of the configuration of the magnetic field of the 'Trimix' multipole trap [227] is shown in Fig. 4.1 *d*. Let us consider this configuration in detail. The basic configuration is created by three internal coils and one coil-baffle, ensuring a stable balance of internal coils. To control the shape of the plasma in the 'Trimix'

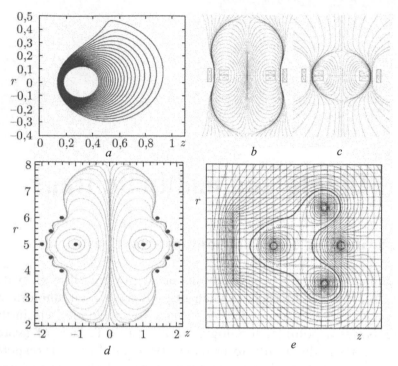

Fig. 4.1. A simple dipole configuration (*a*), a double-dipole trap with elongated (*b*) and compressed (*c*) separatrices [225], a compact dipole with additional conductors [226] (*d*), and the configuration of the 'Trimix' multipole trap [227] (*e*).

trap, a central solenoid (see Fig. 4.1, *d*) or an annular coil placed in its place can also be used.

The possibility of confining the plasma with $\beta \sim 1$ in dipole and multipole configurations allows one to consider the prospects of using these systems for a low-radioactive thermonuclear reactor with D–^{3}He and catalyzed D–D cycles [228].

MHD stability of the plasma in dipole and multipole configurations is achieved both for $\beta \ll 1$ and $\beta \sim 1$, which is demonstrated theoretically and experimentally [220, 229-231].

Plasma confinement in a number of multipole traps was close to the classical one [220]. However, the experiments described in Ref. [220] did not apply to collisionless regimes. Therefore, the question of the role of gradient drift instabilities and the transport caused by them in a collisionless regime is relevant for dipole and multipole systems to the same extent as for other magnetic traps. A theoretical analysis of gradient drift instabilities in the dipole configuration was carried out in [232–234] in the electrostatic approximation. The data on the experimental observation of the drift modes are given in [235].

The boundary of the plasma in multipole traps is the stability boundary, the so-called Ohkawa surface, on which

$$U = \oint_{\psi=\text{const}} \frac{ds}{|B|} = \min, \tag{4.1}$$

where s is the coordinate along the magnetic field, ψ is the magnetic flux function.

Therefore, one of the criteria for optimizing the multipole magnetic configurations is the volume bounded by the Okawa surface.

Because of the azimuthal symmetry, it is sufficient to find the surfaces $\psi = \text{const}$ in the meridian section, i.e., in the (r, z) plane, to construct the magnetic field lines. In cylindrical coordinates, the components of the magnetic field induction vector are related to the magnetic flux function by the relations

$$B_r = -\frac{1}{r}\frac{\partial \psi}{\partial z}, \tag{4.2}$$

$$B_z = \frac{1}{r}\frac{\partial \psi}{\partial r}, \tag{4.3}$$

The magnetic system can be represented in the form of a set of coaxial ring coils consisting of annular turns. The components of the magnetic field of such typical elements [236]

$$B_r = -2 \cdot 10^{-7} I \frac{z}{r\sqrt{(r+a)^2 + (z-b)^2}} \times$$
$$\times \left[K(k^2) - \frac{r^2 + a^2 + (z-b)^2}{(r-a)^2 + (z-b)^2} E(k^2) \right], \tag{4.4}$$

$$B_z = 2 \cdot 10^{-7} I \frac{1}{\sqrt{(r+a)^2 + (z-b)^2}} \times$$
$$\times \left[K(k^2) - \frac{r^2 + a^2 + (z-b)^2}{(r-a)^2 + (z-b)^2} E(k^2) \right], \tag{4.5}$$

where I is the current in the turn, a is the radius of the turn, b is the distance from the plane of the turn to the plane $z = 0$, $k^2 = \dfrac{4ar}{(r+a)^2 + (z-b)^2}$, K and E are complete elliptic integrals.

For given ratios between the radii of all turns and the relations b/a and from (4.4) and (4.5) there follows a criterion for the similarity of magnetic systems:

$$\frac{Ba}{I} = \text{idem.} \qquad (4.6)$$

Therefore, having considered a configuration with specific parameters, it is possible to construct any geometrically similar configurations with the help of condition (4.6). In the calculations, the results of which are given below, the following parameter values were used: the estimated number of turns in the coil $N \approx 50$, the current in the coil $I \approx 200$ A. The currents in the coils were determined taking into account the balance of Ampère forces acting on the internal coils.

The magnetic field lines (surfaces $\psi = $ const) were obtained as a result of numerical integration of the differential equations

$$\frac{dz}{B_z} = \frac{dr}{B_r} = du = \frac{ds}{|B|}, \qquad (4.7)$$

where, for the convenience of numerical calculations, we have introduced the variable u, which, in the process of solving the

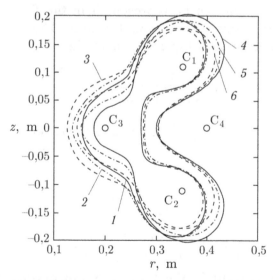

Fig. 4.2. The shape of the Ohkawa boundary for various variants of the magnetic configuration: *1, 4* – without additional coils, *2* – with an additional central ring coil with $N_s = 40$, *3* – with a central solenoid with the number of turns $N_s = 300$, *5* – with an additional central ring coil with $N_s = 120$, *6* – with a central solenoid with $N_s = 600$. For cases *1–3*, $N_3 = N_1$, $N_4 = 1.74N_1$; for *4–5* – $N_3 = 0.8N_1$, $N_4 = N_1$. Coil designation: C_1, C_2 – large internal coils, C_3 – small internal coil, C_4 – pushing coil.

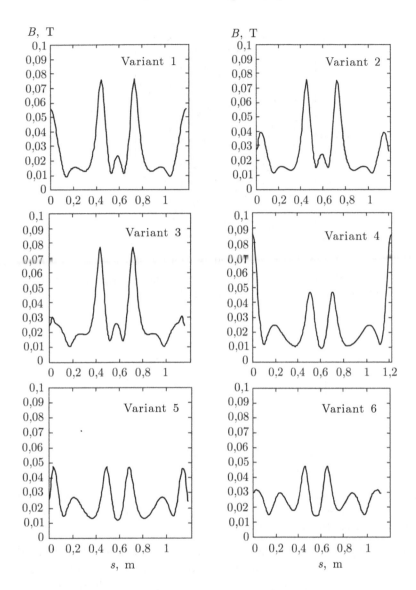

Fig. 4.3. The distribution of the magnetic field along the Ohkawa boundary for the variants presented in Fig. 4.2; s is the coordinate counted along the field line.

equations of the magnetic surface, also allows us to calculate the Ohkawa integral $U = \int du$.

Figure 4.2 and 4.3 show the results of calculations of the magnetic configuration of the Trimix type for the following options: without additional coils, with an additional solenoid, with an additional ring coil.

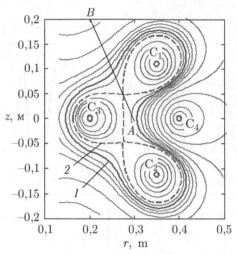

Fig. 4.4. Magnetic field lines (surfaces ψ = const) in a configuration with a central solenoid (option 6). C_1, C_2, C_3, C_4 – see Fig. 4.2.

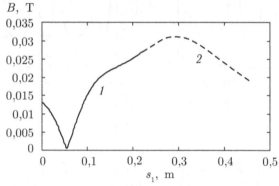

Fig. 4.5. The distribution of the magnetic induction modulus along the line *AB* (see Fig. 4.4): *1* – on the segment *AB*, *2* – on the extension of the segment *AB*; s_1 is the coordinate measured along the ray *AB*.

The shape of the Ohkawa boundary for various variants of the magnetic configuration (without additional coils, with a central solenoid, with an additional central ring coil) is shown in Fig. 4.2. The distribution of the magnetic field along the Ohkawa boundary for these variants is shown in Fig. 4.3. We note that, according to the results of calculations, in the configuration with a ring additional coil the volume of the plasma is somewhat smaller than in the version with a solenoid, but the total current in the ring coil is several times lower than in the solenoid. For one of the variants of the magnetic configuration Fig. 4.4 shows the magnetic surfaces ψ = const. Figure 4.5 shows the distribution of the magnetic field along a given direction.

Let us analyze the inhomogeneity of the magnetic field in the Trimix configuration and a simple dipole in terms of collisionless gradient drift instabilities. This is important for assessing the prospects of reactor systems based on them. As calculations have shown, for the 'Trimix' configuration at the outer boundary of the plasma (on the Ohkawa surface), the parameters $\alpha_B = L_n/L_B$ and $\alpha_R = L_n/R$ strongly vary. The characteristic points of the 'Trimix' configuration are the points of intersection of the boundary with the line AB (see Figs. 4.4, 4.5), where $\alpha_B \sim 0.5$, $\alpha_R \sim -0.1$. The nature of the inhomogeneity of the magnetic field at the outer boundary of the dipole plasma and the 'Trimix' trap is different. At the outer boundary of the dipole plasma, $\alpha_B \approx -1$, $\alpha_R \approx 1$, i.e., the field falls to the periphery of the plasma, and the lines of force are everywhere convex. At the internal boundary of the plasma (near the inner coils) in the configurations in question, $\alpha_B \sim 0.5$, $\alpha_R \approx -0.5$.

For qualitative analysis of gradient drift instabilities, we use the results of Ref. [62]. The analysis leads to the following estimate of the diffusion coefficient in the dipole configuration (for both the outer and inner boundaries of the plasma): $D_\perp \approx 0.1 D_{gyro}$, where $D_{gyro} = \dfrac{\rho_{Ti} k_B T_e}{L_n eB}$. For the 'Trimix' configuration, this value at the outer boundary $D_\perp \approx 0.01 D_{gyro}$, on the inner boundary $D_\perp \approx 0.1 D_{gyro}$. Consequently, the transition from a simple dipole configuration to a multipole configuration of the 'Trimix' type, apparently, gives an average decrease in transport.

Another potential advantage of systems with several internal conductors is a more compact plasma configuration. But this can be judged by the future technical designs of the respective reactors. Analysis of possible technical solutions for levitating coils [237, 238] shows the principal possibility of creating such systems, their functioning in a stationary mode and maintaining the necessary thermal regimes in superconducting windings.

It should be emphasized that the above estimates of the diffusion coefficient correspond to the collisionless regime. Therefore, they are important for promising thermonuclear systems based on dipole and multipole configurations.

It is of interest to distribute a static electric field transverse with respect to magnetic lines of force. In the Oktupol trap (the octupole configuration) [239], the shape of the plasma pinch has a high degree of symmetry, which allows us to speak of a radial electric field E_r. To approximate the intensity E_r, we can propose the

Alternative Fusion Fuels and Systems

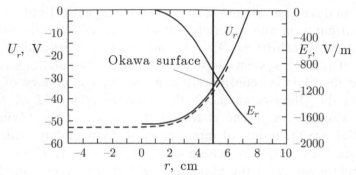

Fig. 4.6. Distributions of the electrostatic field and potential in the central plane of the Oktupol trap. The radial coordinate is measured from the point of zero magnetic field.

following expression: $E_r(r) \approx -\dfrac{kT_e}{e}\left(\dfrac{1}{L_n(r)} + \dfrac{1}{2L_B(r)}\right)$. Figure 4.6 shows the electric field E_r and its potential U_r calculated for the conditions of the magnetic configuration 'Oktupol', and also the experimentally measured distribution of the potential U_{exp} from [240] is shown for comparison.

Consider the possible decrease in transport in the formation of a shear flow under conditions of a collisionless thermonuclear plasma ($T \approx 20$ keV, $B \approx 3$ T, $\beta \sim 1$). at $\gamma_s \gtrsim \gamma$ (γ_s is the shear parameter.) The characteristic growth rate $\gamma \approx 0.1 k_B T/(eBL_n \rho T_i)$. The magnitude of the shear parameter is estimated as $\gamma_s \approx 10 k_B T/(eBa^2)$. Under multipole trap conditions, $\gamma_s \approx \gamma$. In this case, the shear flow gives a decrease in the diffusion coefficient up to five times. Taking into account the shear flow, the value of the diffusion coefficient in thermonuclear conditions can be $D_\perp < 0.1$ m²/s. Then, for a transverse plasma size $a \approx 1$ m, the confinement time can reach $\tau_\perp \approx 5$ s.

Thus, based on multipole traps, it is possible to create a thermonuclear system of acceptable dimensions. In addition, the inhomogeneity factors of the magnetic field, as well as $\beta \approx 1$ in the central part of the plasma column, correspond to the conditions of a lower collisionless transport associated with gradient drift instabilities.

4.2. The main parameters of a D–³He reactor based on a multipole configuration

In this section, the parameters of the D–³He reactor based on the magnetic configuration of the 'Trimix' type, considered in Sec. 4.1, are described. The concept of such a reactor was discussed in [241]. A variant of the fuel cycle with helium-3 production is considered, the relative density of which is $x_{3_{He}} = n_{3_{He}} = 0.3$ corresponds to the maximum ^{3}He production.

Such parameters of the D–^{3}He reactor as the plasma temperature, the size of the plasma column, the induction of the magnetic field, and the required confinement time, have close values for different systems. Therefore, here we will not analyze in detail the energy balance of the plasma in the reactor and the optimization of the system parameters. From the technical point of view, the greatest interest is represented by the magnitude of the currents flowing in the coils of the multipole system, especially in levitating internal coils.

As requirements for the efficiency of the system, we shall establish the values of the volume density of energy release $P_{fus} = 2$ MW/m^3 and the plasma power amplification factor $Q = 20$. The results of the calculations are presented in Table 20.

The value of $P_{fus} = 2$ MW/m^3 corresponds to the minimum requirements, therefore, the magnetic fields in the calculated version are also minimal. Since in this case the induction of the magnetic field reaches 14 T, it is practically impossible to increase the magnetic field, and, consequently, to increase the density of energy release. The sizes of the reactor are relatively large, but their reduction is impractical in terms of total power.

The currents in the levitating coils are about 50 MA. In the design of a reactor with a catalyzed D–D cycle based on a dipole configuration [228], the radius of the inner coil axis is 9 m, the current in it is up to 40 MA. We note that in the analysis of the concept of an internal coil [237] it is shown that for the conditions of its functioning under the conditions of a D–T reactor with fields of level 1 T the minimum current value is 10 MA.

From a technical point of view, the multipole configuration of the D–^{3}He reactor does not look simpler than the compact toroidal systems considered. From the point of view of transport, the multipole magnetic trap has some potential advantages, for example, compared to FRC, associated with the inhomogeneity and curvature of the magnetic field.

Table 20. Parameters of the D–^{3}He reactor based on the Trimix-type multipole configuration

Radius of the axis of the large inner coil, m	$a_1 = 7$
Radius of the axis of the small inner coil, m	$a_3 = 4$
Radius of the axis of the pushing coil, m	$a_4 = 8$
Radius of the axis of the central coil, m	$a_5 = 1.8$
Distance from plane of symmetry $z = 0$ up to large internal coils, m	$b = 2.2$
The total current in the large internal coil, MA	$I_1 = 54$
Total current in the small internal coil, MA	$I_3 = 43$
Total current in the pushing coil, MA	$I_4 = 54$
Total current in the centre coil, MA	$I_5 = 162$
Induction of the magnetic field at the Ohkawa boundary, T	$B = 4.6\text{–}14$
Cyclotron radius of D–^{3}He-protons near the Ohkawa boundary, cm	$\rho_{\text{fus}} = 3\text{–}8$
Density of deuterium, 10^{20} m^{-3}	$n_D = 1.8$
Density of electrons, 10^{20} m^{-3}	$n_e = 2.9$
Plasma temperature, keV	$T_e = T_i = 50$
Beta (maximum density and the minimal field at the Ohkawa boundary)	$\beta = 0.5$
Required energy confinement time, s	$\tau_E > 6$

Magneto-inertial fusion systems

5.1. The concept of magneto–inertial fusion

Magneto–inertial fusion (MIF) is the original direction of inertial controlled thermonuclear fusion (CTF), based on magneto–inertial confinement of hot plasma [242–244]. A new scheme of MIF is the compression of magnetized plasma by powerful laser beams (laser-driven MIF) or high-speed plasma jets (plasma jet driven MIF) [55, 245–247]. In MIF, as in magnetized target fusion (MTF), a spherical or cylindrical plasma configuration is compressed in an external seeding magnetic field. In recent years, the very promising direction of CTF [249] has been actively developing, where a cylindrical metal liner is compressed under the action of a powerful Z-pinch, and a thermonuclear process in a hot compressed cord is initiated by a femtosecond laser.

MTF is considered as a subclass of MIF with the use of gaseous, liquid and solid shells and impactors (liners). MIF systems unite a wide range of thermonuclear installations where the compression of magnetic flux is possible with different types of shells, including gas hammers, liners formed by the fusion of high-speed (Mach number $M > 5$) plasma jets, lasers, heavy ion beams, etc. The leading MTF scientific groups are in China, Russia, the United States and Japan.

The research of MTF is stimulated by the significant progress achieved in recent years: experiments on the compression of the magnetic flux by laser beams and plasma jets, the creation of magnetic systems with a high ratio of plasma pressure to magnetic field pressure and the use of modern laser installations, plasma guns and liners. These achievements open up new opportunities for the development of efficient methods of obtaining energy from the plasma.

The idea of separate compression and initiation of thermonuclear reactions by another source was first formulated in the mid-1980s [250]. The scheme of additional heating due to a strong shock wave moving from the peripheral layers of the compressed target was discussed in [251]. Later such a scheme was investigated using mathematical modeling methods. In fact, the MIF began with the study of pinches, which, along with stellarators and magnetic traps, were among the very first directions in the study of thermonuclear fusion and methods for generating a pulsed magnetic field. Like other concepts of pulsed power engineering, MIF aims at plasma implosion to high temperature and its confinement for a time sufficient to produce fusion energy. Under the MIF conditions, a plasma with a high energy density can be formed [252] due to heating by fusion products [253].

The diagram below shows possible variants of powerful energy sources (drivers), magnetic configurations as targets and liners for systems and promising directions of magnetic–inertial fusion.

Let us emphasize once again that MIF unites all MTF concepts, including magnetic hydrodynamic reduction [254], magnetic compression of liners [255], pinch [256]. The MIF concept includes generation schemes as well as plasma heating: laser heating, Z-pinch, plasma focus, liner schemes with plasma compression by fields or high-speed impactors, 'exploding wires', laser-heated plasma inside the solenoid [257], antiprobotron (cusp) [258], cryogenic and stabilizing Z-pinch [259], θ-pinch and spindle-shaped anti-mirror

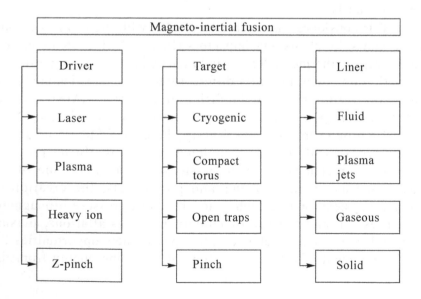

geometries [260] and combined systems [261]. The specificity of MIF is that this approach requires a liner [262] to compress and heat a magnetized plasma (target), for example a compact torus [263].

Various variants of an explosive pusher have been considered previously, allowing to preserve the magnetic flux: a metallic liner [264], a partially evaporated liner [265], a gaseous liner [266], compressible liquid shells [267], an implosion in the pusher [268], a fast ignition of thermonuclear fuel. Of particular interest is the very promising direction of controlled thermonuclear fusion developed at Moscow State University by V. Voronchev et al. [55, 248], where a cylindrical metal liner is compressed under the action of a powerful Z-pinch, and a thermonuclear process in a hot compressed cord is initiated by a femtosecond laser. This new direction, developed at Moscow State University, has received active further development in new works of scientists from the Sandia laboratory.

It was shown in [269] that the ignition of thermonuclear fuel is greatly simplified in the case of a magnetized plasma, where the optical thickness of the fuel is $\rho R < 1$ g·cm^{-2}, and the product of the magnetic field by the radius of the target is $BR > 60$ T·cm. The prospects of the reactors based on fusion with magnetized targets and the problems of power plants based on a pinch with an inverted field were considered in [270].

High luminosity values correspond to lasers and ion beams as drivers, used, for example, in plants [271] and [272]. For typical scenarios of inertial thermonuclear fusion, magnetic fields from 1000 to 10 000 T are required, which are generated by large currents flowing in small outer coils surrounding the target. The ignition of the reaction in this case is possible at a low implosion rate due to the magnetization of the target.

Plasma jets and various liners provide only average and low values of β [273–275]. The gain factor for economically feasible energy production can be much lower than for the laser driver of inertial fusion. This also includes solid and liquid shells, non-cryogenic gas targets and high-performance cheap drivers, for example [276, 277]. Low values of the coefficient β were obtained in experiments [147, 278–280].

In other words, the classical pinches produced a low plasma temperature, which was completely inadequate for CTF, and also with a large number of instabilities. Therefore, currently drivers with plasma jets (plasma liner) and laser beams (laser driver) and combined circuits, for example Z-pinch with ultrafast laser ignition

[55, 248, 249] are most attractive. In the case of a target compression by plasma jets or a laser beam, the configuration is formed inside the combustion chamber under the action of an external magnetic field (solenoid).

In experiments with plasma liners formed by fusion of plasma jets, their ability to evolve to spherical symmetry is tested [281, 282]. An acoustic driver [283] – pneumatic impact pistons with an energy of 500 kJ and a pulse duration of 80 s, capable of producing convergent shock waves in a plasma compression fluid up to 10^{20} cm^{-3} and a temperature of 10 keV was developed [283]. At the moment in Japan, the compression of the electromagnetic flow has reached a record value of the magnetic field of 700 T [284]. In Russia, in addition to the above-mentioned works [55, 248] on the creation of Z-pinch with superfast laser ignition, experimental and theoretical work is carried out on three-dimensional compression of quasi-spherical wire liners [285], on compression of an external longitudinal magnetic field as a result of electrodynamic implosion of a cylindrical multiwire liner [286], modelling the motion of a plate liner [287] and the interaction of high-power laser and plasma beams with a magnetized plasma target [288–290].

The following scheme is considered: at the initial moment of time (before switching on the driver) in the working body of the target of inertial thermonuclear fusion, an initial magnetic flux is formed. Further, the thermonuclear target is irradiated on all sides by intense radiation from a multichannel laser (with an emission intensity of 10^{15}–10^{17} W·cm^{-2}). As a result, the formation of a magnetized plasma is possible in compressed fuel. The magnetic field is compressed and creates a dynamic pressure on the target shell from the inside, heating it up to ultrahigh temperatures. This increases the density of the magnetic flux, creating an extremely strong magnetic field.

The compression of the magnetized target and the magnetic flux generated by the external coils for the case of direct laser irradiation proceeds according to the following scheme. The seed magnetic field is created by a current flowing in the rings, the value of which on the OMEGA laser [271] is 80 kA [291]. For experiments with a magnetic field, a generator was created, which is a compact system with accumulators, a spark gap and a transmission line, which is inserted and removed after each shot to change the coils. The driver has about 100 J of energy, the pulse duration is 400 ns, which is enough to create a magnetic field of the order of 10 T in the centre. The coils are made of foil with a thickness of 0.3 mm and a width of 0.8 mm, their diameter is about 7 mm and the distance between the rings is about 8 mm.

The requirements for target ignition in inertial thermonuclear fusion can be significantly relaxed if the plasma is compressed simultaneously with the trapped magnetic flux [292, 293]. A superstrong longitudinal magnetic field isolates the hot core from the cold walls, compressing it and holding thermonuclear α-particles. Two approaches to compressing the magnetic flux by a pulse power source [294] and laser ablation [295] demonstrated their efficiency, generating maximum fields of 42 and 36 MG, respectively. The laser apparatus [271] demonstrated experimentally an increase in the thermonuclear yield by compression of an external magnetic field introduced into thermonuclear fuel. In the case of a spherical implosion of a solenoidal magnetic field with open lines of force, an increase in the neutron yield by 30% was obtained. The deflection of fast protons measured a compressed magnetic field for spherical implosion 23 MG. If a magnetic field with closed lines of force can be introduced into the target plasma, we can expect an increase in the neutron yield by a factor of 2–4. A full two-dimensional radiation-hydrodynamic simulation of the effect of highly compressed magnetic fields on the ignition of spherical capsules with cryogenic fuel was carried out in [296], it is proposed to carry out experiments with a seed magnetic field at a NIF facility.

At the present time, theoretical and experimental studies of spherical and cylindrical implosion of the plasmoid are being actively carried out, where plasma guns (railguns) act as the driver, and a plasma liner that compresses the formed magnetic configuration is obtained by uniform fusion of supersonic jets.

Magnetic reduction of the target by high-speed plasma jets is possible in the open trap scheme [297, 298], pinch [299, 300], spheromak and reversed magnetic configuration [301]. In the latter case, it is meant to create a magnetized plasma in the formation chamber and transport it to the combustion chamber. That is, unlike the scheme with the laser driver, we have a preformed magnetic configuration, which is then compressed by accelerated beams.

For ZEBRA setup parameters [302], the implosion of the liner system in the Z-pinch configuration was simulated, where the target is effectively heated first by shock waves, and then – under adiabatic compression. The thickness of the liner is regulated by radial current transfer and subsequent amplification of the current in the target. In the case of a target with a high atomic number, the shock wave does not play a significant role in heating, and the compression ratio required to heat a plasma with a low atomic number is significantly reduced.

In [303], scaling is presented for the thermal pressure and stagnation time during the implosion of a spherical plasma liner, formed by the fusion of several plasma jets. It is established that the thermal pressure is scaled linearly with the number of jets, the initial value of the jet density and the Mach number, quadratic with the initial radius of the jet and the velocity and inversely proportional to the initial jet length and the square of the radius of the chamber. The obtained results formed the basis of PLX (Plasma Liner eXperiment) [304].

Additional heating and current maintenance in a dense plasma sphere can also be applied by using laser or electron beams to create a magnetic field frozen into the plasma target [305]. In a pulse system of this type, it is possible to achieve a mega-Gaussian magnetic field, and its freezing into the plasma occurs in less time than the time of the driver's action. In most variants the main problem is the transfer of the driver's energy to the plasma, the conversion of the energy of the external source into the kinetic energy of the plasma. Therefore, lasers with a high pulse energy ($> 10^{15}$ W·cm^{-2}) and a magnetic confinement system for a magnetized plasma with a higher β coefficient are chosen. Such systems include a compact torus, in particular an inverse magnetic configuration of FRC (field-reversed configuration) [306, 307], for which stable formation and magnetic compression is demonstrated. At the IPA (Inductive Plasma Accelerator) for deuterium plasmoids, a velocity of 600 km·s^{-1} was experimentally obtained [147]. The inductive compression of the liner for $B_z > 1$ Mbar fields [308] was demonstrated, as well as the successful compression of FRC with a xenon plasma liner [309] and Z-pinch with beryllium [310] and fiber liners [256]. For the convergence of three plasma jets [277], a compact (1 cm long transverse scale), a high-density (1.2×10^{17} cm^{-3}), low-temperature (1.5 eV) region with a high pressure (70 kPa) was detected. HyperV Technologies Corporation developed and manufactured for the MCX installation (Maryland Centrifugal eXperiment) a coaxial plasma gun that generates plasma jets with a mass of 160 µg at a speed of 90 km·s^{-1} and a plasma density of more than 10^{14} cm^{-3} [278].

In the last ten years, significant progress has been made in the study of MIF concepts. Let us dwell on some of the difficulties and issues that were encountered in the study of this direction. The main task of MIF is to determine how the generated magnetic field can facilitate ignition or increase the thermonuclear yield for various schemes of inertial fusion. There are several issues that need to be addressed: can plasma targets be formed and then compressed and heated to thermonuclear temperatures; what are the transport mechanisms for the loss of particles, energy and flow; what is the stability of the initial target configurations. Each of these issues includes components of engineering and fundamental science that are overlapped by the interests of inertial thermonuclear fusion and other fields of knowledge, including materials science at high pressures and astrophysics. Superstrong magnetic fields can change the properties of matter in a difficult-to-predict direction, and the Rayleigh–Taylor instability is a key issue when considering liner processes.

The MIF systems tend to a larger energy output and a lower frequency of repeatability. It is possible to use liquid walls, which are also relevant for other targets and drivers, especially for heavy-ion beams of thermonuclear fusion. The use of MIF for energy production has not been considered as widely as the traditional systems of magnetic or inertial thermonuclear fusion, and therefore the parameters of inertial fusion are less clearly defined. The thermonuclear yield in the gigajoule range gives an advantage to MIF systems with a lower pulse repetition rate compared to conventional inertial fusion, although the concept of a plasma liner is intermediate and is designed to yield well below 1 GJ and a repeatability of about 1 Hz. At present, several MIF options are being investigated: 1) pulsed compression with a circulating liquid metal; 2) the concept of a reactor with a fast liner, using a liquid blanket; 3) drivers at a distance – plasma jets, lasers, ion beams and electron beams.

The dynamics of plasma jets in MIF has been studied for the last decade, in particular, because the version of plasma jets [311] bypasses the difficult engineering and technological problem of excessively rapid combustion of the electrode for solid impactors and slow implosion velocity in the case of liquid or metal liners for MIF. In addition, outside the plasma target it is possible to form a kind of internal magnetic shell (IMS), i.e., the inner layer of the liner. One of the main functions of IMS is the thermal insulation of the target plasma from the liner, taking into account the interaction of the products of fusion with the liner. The most important question is: will the products of the thermonuclear reaction sufficiently warm up the IMS and the plasma liner to produce a significant amount of energy? The target parameters with IMS and the plasma liner are given in [312].

The latest results of numerical simulation [313] predict the possibility of ignition and a high value of the coefficient β when compressing the flow for the next generation of high-current installations with currents of the **multimegaampere** range. The key issue of obtaining a high value of β of the gain of thermonuclear energy for all variants of this approach is the propagation of thermonuclear burning through superstrong magnetic fields from the hot region to the cold fuel.

Plasma liners offer opportunities such as the use of composite jets/liners with DT-fuel and exclude the need for separate target formation, liner profiling and delivery of additional fuel to enhance combustion and fusion output. In comparison with conventional inertial fusion systems in MIF circuits with a laser driver and a plasma liner, heating is possible when implosion of liners, lasers or particle beams with reduced power and intensity is used. The development of concepts can be very fast, because the necessary scientific research, including the physics of plasma combustion, does not require new expensive installations. In addition, the successful implementation of the concept of a reactor with a liquid wall will also reduce the costs of research in materials science.

MIF can be considered as a broader class of inertial thermonuclear fusion, which is characterized by relaxed requirements for drivers and targets, although it is complicated by the imposition of a magnetic field. Advantages of MIF with a laser driver and a plasma liner consist in the possibility of compressing the original magnetic field to several hundred and even thousands of Tesla, in relatively large values of DT plasma concentrations and thermal insulation by a magnetic field. In addition, additional adiabatic compression by laser beams or plasma jets of a preformed low-temperature plasma and a 'frozen' magnetic field in it is carried out in the MIF. Such systems can be used to carry out experiments with low-radioactive or neutron-free fuels [64, 248, 314], as well as for the diagnosis and testing of various materials [315, 316]. Physics of MIF qualitatively differs from the physics of inertial thermonuclear fusion, so the process of finding the optimal solution for a target with a high energy yield and parameters of plasma jets may require other approaches and analysis.

To date, quasi-stationary systems have been studied in more detail, and in this paper both quasi-stationary and pulsed installations are considered. The advantage of such energy systems is their compactness, smaller losses, and micro-stability does not have time to develop. The hydrodynamic instabilities are due to the inhomogeneity of the flows. The plasma–wall problem is not so relevant. The problem may be impurities, but not as much as in systems with metal and liquid liners. Compared with inertial methods of thermonuclear fusion (ICF) – the pulse duration and confinement is longer. Also, the advantages of such systems and configurations are: high frequency of fuel supply, quasi-stationary (provided by lasers or additional heating systems), high burn-out rate – less fuel required, more energy release. The novelty and peculiarity of this task is to find the optimal compression methods: the composition of the liner and the configuration of external drivers, the definition of perspective compression modes over time, the requirement for a technique that provides this compression.

Two approaches to compressing the magnetic flux by a pulse power source [294] and laser ablation [295] demonstrated their efficiency, generating maximum fields of 42 and 36 MGS, respectively. The laser apparatus [271] demonstrated experimentally an increase in the thermonuclear yield by compression of an external magnetic field introduced into thermonuclear fuel. In the case of a spherical implosion of a solenoidal magnetic field with open lines of force, an increase in the neutron yield by 30% was obtained. The deflection of fast protons measured a compressed magnetic field for spherical implosion 23 MG. If a magnetic field with closed lines of force can be introduced into the target plasma, we can expect an increase in the neutron yield by a factor of 2–4. A full two-dimensional radiation-hydrodynamic simulation of the effect of highly compressed magnetic fields on the ignition of spherical capsules with cryogenic fuel was carried out in [296], it is proposed to carry out experiments with a seed magnetic field at a NIF facility.

In experiments with plasma liners formed by fusion of plasma jets, their ability to evolve to spherical symmetry is tested [277, 278]. An acoustic driver [317] – pneumatic impact pistons with an energy of 500 kJ and a pulse duration of 80 s, capable of producing convergent shock waves in a liquid for compressing the plasma to 10^{20} cm^{-3} and a temperature of 10 keV was developed [317]. At the moment in Japan the compression of the electromagnetic flow has reached a record value of the magnetic field of 700 T [319]. In Russia, work is underway to create a Z-pinch with ultrafast laser ignition [55], experimental and theoretical work on three-dimensional compression of quasispherical wire liners [319], compression of an external longitudinal magnetic field as a result of electrodynamic implosion of a cylindrical multiwire liner [320] the motion of a plate liner [321] and the interaction of high-power laser and plasma beams with a magnetized plasma target [288, 290].

MIF, often referred to as MTF, is a concept alternative to MCF and ICF. The trajectories of the motion of thermonuclear products (alpha particles and protons) and the heating problem in MIF/MTF – conditions are an important point in the study of this topic. The dynamics of plasma jets in MTF [245, 253] has been the object of study for scientists around the world during the last decade in particular because the version of plasma jets bypasses the difficult engineering and technological problem of excessively rapid combustion of the electrode for solid impactors (liners) and slow implosion velocity in the case of liquid or metal liners for MTF. However, it should be noted that the proposed study is applicable for all versions of MIF/MTF and any thermonuclear fuel cycle. Much of this work, of course, with modification for a specific configuration of the magnetic field, can be used for traditional concepts of CTF.

5.2. Rocket and isoentropic models of laser reduction of a magnetized spherical target

With reference to our specific task, we can use the term MLF – magneto-laser fusion, which uses targets standard for ICF. The dynamics of plasma in the interaction of a target with powerful laser beams in MIF is an object of study for scientists throughout the world during the last decade in particular because the laser driver bypasses the difficult engineering and technological problem of the excessively rapid combustion of the electrode for solid impactors (liners) and slow implosion velocity (shock compression) in the case of liquid or metal liners. Note that the proposed study is applicable for all versions of MIF and any thermonuclear fuel cycle.

The presence of a magnetic field in the plasma of the target significantly reduces the losses associated with thermal conductivity, which allows, first, to use a relatively slow uniform compression and heating of the plasma to the conditions of fusion and, secondly, weaken the requirements for the energy of the laser pulses needed for crushing. The main advantages of this approach are lower implosion rates, energy costs and a decrease in the electronic thermal conductivity (in comparison with the ICF) due to the use of an external seeding magnetic field, and also small plasma confinement times (in comparison with the MCF).

The dynamics of a plasma located in a seed magnetic field with an antiprobe configuration (cusp) squeezed by laser beams is studied. Until recently, in the studies on MCF, the cusp configuration of the magnetic field was considered not very successful for confinement of charged particles (although work on this system is still ongoing [258, 322]). Nevertheless, this configuration can be attractive for MIF, where superstrong magnetic fields are excited, and the necessary confinement time of particles can be sufficiently small.

The scheme for MIF with a laser driver works as follows: a pulsed solenoid creates an initial magnetic flux (in our case, the field is generated by ring currents flowing in opposite directions). After the formation of the seed magnetic field, the laser driver is turned on, as a result of which the target is compressed by several laser beams – the target is compressed in the magnetic field of the antiprobe configuration. The magnetic field is compressed and creates a dynamic pressure on the target shell from the inside, heating it up to ultrahigh temperatures. This increases the density of the magnetic flux, forming an ultrastrong magnetic field.

The decrease in the losses due to thermal conductivity in the hot core of a magnetized plasma leads to an increase in core temperature at lower compression rates than those required in ICF. Thus, it is possible to achieve combustion with a high energy yield. To model the interaction of laser radiation with a spherical target, a compact antiprobe magnetic configuration, or cusp (Figs. 5.1 and 5.2), in which a strong macroscopic magnetic field is created, is chosen. The arrows show the lines of force of the magnetic field, and the dots and crosses indicate the direction of the current in the coils. The area occupied by the plasma is painted over. O – point of zero magnetic field, δ_0 – width of axial (point) cusp, δ_L – width of annular (linear) cusp.

To realize the magneto-inertial regime, it is necessary to generate a sufficiently strong seed field. A model of a high-current generator of a seed magnetic field was developed and experimentally tested in [291]. The intensity of the magnetic field excited by the generator reached in the experiments was 10 MG.

In contrast to [291], where a seed magnetic field generator was used, using two turns with a current flowing in the same direction, we suggest turns with current flowing in opposite directions to be used to generate a seed magnetic field of the antiprobe configuration

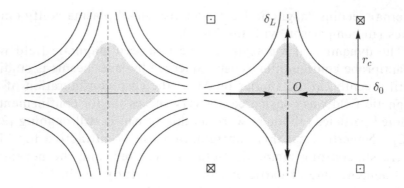

Fig. 5.1. Magnetic configuration of the cusp. The magnetic field lines and the region occupied by the plasma are shown.

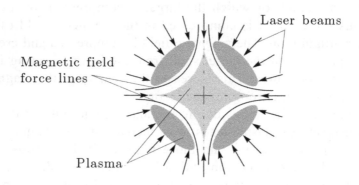

Fig. 5.2. Spherical magnetic cusp configuration – compression of magnetic flux and plasma by laser beams.

(Fig. 5.3). During the implosion, the magnetic flux is frozen inside the hot region into the plasma. When the plasma is compressed, the magnetic flux also contracts, and the intensity of the magnetic field reaches several thousand Tesla.

The schemes proposed earlier dealt mainly with compressible liners and electron beams [246, 323]. Attempts to analyze the behaviour of a thermonuclear magnetized target under the action of high-power laser beams have not been made, and such works have appeared only recently due to the progress of these systems [296]. One of the issues discussed in this paper is the generation and excitation of spontaneous magnetic fields in a magnetized plasma under the action of a laser driver. The peculiarity of this problem, in contrast to a laser jet??? or interaction with a plasma of high-

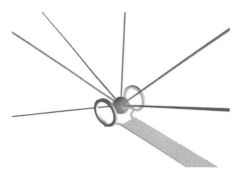

Fig. 5.3. Scheme of spherical implosion of a magnetized target with uniform compression by laser beams.

intensity lasers, is the presence of a strong magnetic field. The approach proposed in [258, 291] opens up prospects for using this method of energy generation. A mathematical model of the interaction of laser beams with a high pulse energy with a plasma target in a seed magnetic field was proposed in [288].

Next, consider the compression of the shell in the framework of the 'rocket' model. Since we investigate the compression of a target under the action of laser beams, it is necessary to take into account an important phenomenon, such as target ablation. Under the action of laser radiation, the outer part of the shell evaporates. The flow of plasma to the outside results in an inverse recoil effect directed into the interior of the target and leading to compression of the inner part of the target. The initial stage of compression is conveniently described in the framework of the 'rocket' model [324]. We assume that the plasma inside the target can be represented by an equivalent sphere of radius r, which is held by a spherical plastic shell of mass m and radius r. As shown in [325], for a symmetric cusp??? configuration, the correction coefficient connecting its characteristic radius with the radius of the equivalent sphere $r = gr_{\mathrm{cusp}}$ is equal to $g = 0.6$. The use of the laws of conservation of mass and energy-momentum leads to the following expression for the pressure on the inner part of the target [326]:

$$p = 2\dot{m}v_{\exp},\qquad(5.1)$$

where $v_{\exp}$ is the velocity of the outward gas flow. Integrating equation (5.1), we obtain the basic equation of the 'rocket' model

$$v(t) = v_{\exp} \ln \frac{m_0}{m}. \tag{5.2}$$

Thus, for the velocity of implosion, we can write

$$\frac{dr}{dt} = v_{\text{imp}} = v_{\exp} \ln \frac{m_0}{m} = \frac{p}{\dot{m}} \ln \frac{m_0}{m}. \tag{5.3}$$

Using the equation of state of an ideal gas

$$NkT = pV, \tag{5.4}$$

the expression for the volume of a spherical target

$$V = \frac{4}{3} \pi r^3, \tag{5.5}$$

the expression for the target mass

$$m = m_i N, \tag{5.6}$$

equation (5.3) can be rewritten in the form

$$\frac{dr}{dt} \frac{dN}{dt} = \frac{3kTN}{4\pi r^3 m_i} \ln \frac{N_0}{N}, \tag{5.7}$$

where T is the plasma temperature, k is the Boltzmann constant, N is the number of particles in the cusp0, and m_i is the mass of the ion. For simplicity, we assume that the temperature is uniformly distributed inside the target. The loss of plasma particles from a magnetic trap with a cusp configuration can be estimated from the dependence [327]

$$\frac{dN}{dt} = -\frac{16nkTrc}{eB}, \tag{5.8}$$

where $n = N/V$ is the plasma density, B is the magnetic field strength, e and c are the ion charge and the speed of light, respectively.

Compression of the magnetic field is determined by the formula [291]

$$B \approx B_0 \left(\frac{r_0}{r}\right)^{2(1-1/\text{Re}_m)}, \tag{5.9}$$

where $\text{Re}_m = \mu\sigma v_{imp}r$ is the magnetic Reynolds number that determines the ratio of the rate of reduction to the rate of magnetic diffusion, μ, σ and r are the magnetic permeability, conductivity and spatial scale of implosion, respectively.

Further, we set $\mu \gg 1$. Using the relations (5.5) and (5.9), equation (5.8) can be represented in the form

$$\frac{dN}{dt} = -\frac{N}{\tau_{loss}},\tag{5.10}$$

$$\tau_{loss} = \frac{\pi}{12}\omega_{ci}\frac{r_0^2}{v_i^2},\tag{5.11}$$

where $\omega_{ci} = eB_0/(m_ic)$ is the ion synchrotron frequency, $r_0 = r(t = t_0)$ is the initial radius of the sphere, $v_i = (kT/m_i)^{1/2}$ is the thermal velocity of the ions. The solution of equation (5.7) has the form

$$\tilde{N} = \frac{N}{N_0} = \exp\left(-\frac{t}{\tau_{loss}}\right),\tag{5.12}$$

where $N_0 = N(t = t_0)$ is the initial number of plasma particles.

Substituting the solution (5.12) into equation (5.7), we obtain an equation describing the compression of the shell:

$$\frac{da}{d\tau} = -\frac{\tau}{a^3},\tag{5.13}$$

where the dimensionless variables $a = r/r_0$, $\tau = t/t_{comp}$ and the following notation: $\tau_{comp} = (4\pi p_0/m)^{-1/2}$, $\alpha = \tau_{comp}/\tau_{loss}$, p_0 is the initial value of the pressure on the shell. The solution of the equation has the form

$$a = 1 - \frac{\sqrt{\tau}}{2}.\tag{5.14}$$

Obviously, the solution obtained describes only the initial stages of compression ($r \approx r_0$, $\tau \ll 4$), since at large times the target radius becomes negative. We also calculate the distribution of concentration and pressure inside the target. According to the equations (5.4), (5.5), (5.12), (5.13) we obtain

$$n = \frac{p}{T} = \frac{N}{V} = n_0\left(1 - \frac{\sqrt{\tau}}{2}\right)^{-3}\exp\left(-\frac{\tau}{\tau_{loss}}\right),\tag{5.15}$$

where $n_0 = 3N_0 / (4\pi r_0^3)$.

Isoentropic compression was studied for the first time in [328, 329], where it was shown that this regime leads to more efficient consumption of the driver energy to obtain a high compression ratio. Such a compression can be provided by choosing a special profile of the intensity distribution of the laser pulse in time [330]. Let us now consider a model for isoentropic compression, which makes it possible to investigate later compression stages ($r \ll r_0$). The implosion of the shell is described by the equation [326]

$$\frac{d^2 r}{dt^2} = \frac{4\pi r^2 p}{m},$$ (5.16)

where p is the pressure on the shell. Using equations (5.4), (5.5), equation (5.1) can be rewritten in the form

$$\frac{d^2 r}{dt^2} = \frac{3NkT}{rm}.$$ (5.17)

The losses of plasma particles are described by formula (5.12). Substituting the solution of (5.12) into equation (5.17), we obtain an equation describing the compression of the shell

$$\frac{d^2 a}{d\tau^2} = \frac{1}{2}\exp[-\alpha\tau].$$ (5.18)

In the solution obtained, there is no singularity at the time of maximum compression, in contrast to the rocket model, which allows the later stages of compression. Equations (5.15) can also be used to find the density and pressure distributions inside the target.

Next, we numerically solve the system of equations (5.12) and (5.18) for the initial values $a(t = 0) = 1$, $\dot{a}(t = 0) = 0$ for different values of α. The system solutions are shown in Fig. 5.4. The values for different parameters of the magnetized plasma are given in Table. 21. The results of the magnetic-inertial fusion experiment with a laser driver [291] correspond to the lower part of the table.

Note that in our model $t = 0$ corresponds to the minimum radius, that is, to the maximum compression, and not to the initial instant of time. The origin is taken for a negative value of t (see Fig. 5.4), at which we have an initial implosion velocity directed toward the centre of the target. As a result, in our model, as in the Robson

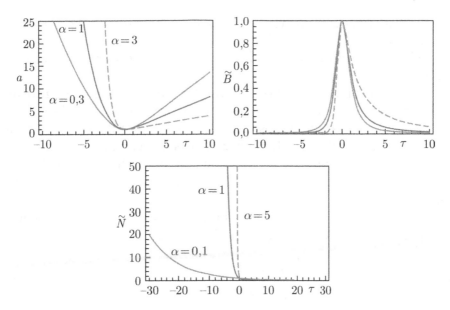

Fig. 5.4. Distributions of dimensionless radius a, magnetic field $\tilde{B}$ and number of particles $\tilde{N}$ of a quasispherical plasma in a magnetic field of opposing ring currents as a function of the dimensionless time τ for the cusp magnetic configuration for different values of α. Zero on the horizontal axis corresponds to the moment of the target compression by lasers.

Table 21. Plasma parameters for different values

$B = 10$ T, $T = 7$ keV		$B = 1000$ T, $T = 20$ keV		$T = 5$ keV, $r_0 = 0.0005$ m	
r_0, m	α	r_0, m	α	B, T	α
10^{-2}	0.44	10^{-3}	0.12	10	6.24
$5\cdot10^{-3}$	0.87	$5\cdot10^{-4}$	0.25	50	1.24
10^{-3}	4.37	10^{-4}	1.25	100	0.624
$B = 10$ T, $T = 1$ keV		$T = 7$ keV, $r_0 = 0.00043$ m		$B = 150$ T, $T = 5$ keV	
r_0, m	α	α	B, T	r_0, m	α
10^{-3}	0.62	0.03	3000	10^{-3}	0.21
$4.3\cdot10^{-4}$	1.45	0.20	500	$5\cdot10^{-4}$	0.42
10^{-4}	6.24	1.01	100	$4.3\cdot10^{-4}$	0.48

model [325]), the target shrinks, reaches a certain minimum radius, and then expands due to internal pressure.

Let us turn to the discussion of the magnetized target MIF in a thermonuclear reactor. It is proposed to use, as a target, cusp magnetic configurations or similar systems with high β (the ratio of the plasma pressure to the magnetic field pressure) and the current flowing in the small outer coils surrounding the target to generate a seed magnetic field whose intensity increases as the target is compressed.

There are three basic requirements for ignition and burning capsules:

1) the fuel temperature should be higher than the ignition temperature (for an equicomponent D–T mixture it is 4.3 keV);

2) it is necessary to enclose a certain fraction of the thermonuclear energy in the fuel target for subsequent self-heating;

3) the fuel should be kept as long as possible in order for the output fusion energy to exceed the driver's energy embedded in the capsule.

The total thermonuclear power produced per cycle per unit volume of plasma is, in general, equal to (in MW/m³)

$$P_f = \left(\frac{\beta B^2}{2\mu_0}\right)^2 \frac{<\sigma v> E_f}{T^2(z_1 + z_2 + 2)}, \tag{5.19}$$

where B is the external magnetic field, μ_0 is the magnetic constant, $\langle\sigma v\rangle$ is the rate of the thermonuclear reaction averaged over the Maxwellian distribution, E_f is the fusion energy, T is the plasma temperature, z_1, z_2 are the charges of the reacting particles.

The change in the neutron energy for a D–T mixture can be described by the formula (in MW)

$$P_n = \left(n_D n_T \langle\sigma v\rangle_{DT} E_n^{DT} + \chi \frac{n_D n_D}{2} \langle\sigma v\rangle_{DD} E_n^{DD}\right) \frac{4}{3}\pi r^3, \tag{5.20}$$

where n_D, n_T are the concentrations of deuterium and tritium respectively, $E_n^{DT} = 14.07\,\text{MeV}$ and $E_n^{DD} = 2.45\,\text{MeV}$ are the neutron production energies in the D–T and D–D reactions, χ is the fraction of the reacted tritium nuclei.

To calculate the velocities of the main thermonuclear reactions (in m³/s) a simplified formula [331] can be used in a narrow temperature range (in keV) typical of MIF, which gives less than 1% error in comparison with the data [81, 332].

The energy of the alpha particles remaining in the combustion zone, normalized to the entire energy of the produced alpha particles, has the form [333]:

$$E_{\text{H}} = \begin{cases} \left(\dfrac{16}{15} + \dfrac{8}{15}\ln 2\right)\tau - \dfrac{28}{36}\tau^2 \text{ at } \tau \leqslant \dfrac{1}{2}, \\[2ex] \left(\dfrac{248}{450} - \dfrac{8}{15}\ln \tau\right)\tau + \dfrac{1}{9}\tau^2 + \dfrac{1}{9}\tau - \dfrac{1}{18\tau^2} + \dfrac{1}{450\tau^4} \text{ at } \dfrac{1}{2} \leqslant \tau \leqslant 1, \\[2ex] 1 - \dfrac{1}{3\tau} + \dfrac{1}{18\tau^2} - \dfrac{1}{450\tau^4} \text{ at } 1 \leqslant \tau, \end{cases} \quad (5.21)$$

where $\tau = r/r_\alpha$, r_α is the deceleration length of the alpha particle.

As an effective magnetic field we considered a value corresponding to the equality of the plasma pressure and the magnetic pressure. To determine the minimum requirements for the magnitude of the magnetic field at the final stage of compression, numerical calculations of the evolution of the balance of particles and energy in time were made. As the results of [334] show, for the most promising regimes, magnetic fields are required at a level $B > 300$ T and more.

Giant magnetic fields are required to effectively compress and isolate hot plasma. Recent experiments demonstrate restrained optimism in the generation of magnetic fields. In experiments with compression, it was possible to reach a magnetic field strength of 30–40 MG. Apparently, the optimization of the experiment will make it possible to achieve substantially higher values of the magnetic fields.

Without considering here in detail the requirements for the symmetry of the compression of a spherical fuel capsule, which are very stringent, we note that the data obtained is sufficient for analyzing the MIF system as a reactor or particle source. In the future, problems connected with the introduction of plasma targets into the reactor zone, with cleaning from impurities and with a regime of frequency impulse loads, should be considered in detail. This will make it possible to draw more substantiated conclusions about the competitiveness of this direction. The results of parametric analysis can be used as input data for engineering physics research.

The use of magnetically inertial confinement of a magnetized target with subsequent compression significantly reduces the power requirements of the driver that compresses the target. The presented scheme of laser reduction of a magnetized target with an a cusp configuration of the magnetic field can be used to develop a magnetically inertial fusion reactor with a laser driver.

5.3. The configurations of the target and the liner formed by supersonic gas and plasma jets

A distinctive feature of this problem is the presence of external magnetic coils and, as a consequence, a sufficiently strong priming magnetic field. A magnetic–inertial approach is considered, more specifically a subset of magnetic–inertial thermonuclear fusion for the realization of a thermonuclear reaction, which has the advantages of both concepts of CTF – high values of the energy density of inertial thermonuclear fusion and thermal isolation of plasma by a magnetic field typical for magnetic thermonuclear fusion. The decrease in losses due to the thermal conductivity in the hot core of the exploding magnetized plasma (target) leads to an increase in core temperature at lower compression rates required by traditional ICF. In this case, it is possible to achieve combustion with a high energy yield. For modelling, a compact pulsed magnetic system is created that creates a macroscopic magnetic field in a spherical target (a compact torus) and compares the mechanisms of formation of a plasma liner by high-speed plasma jets.

In the process of MIF, the striker (liner) is used to compress the plasma to the conditions of fusion and inertial confinement of the burning plasma to obtain the necessary energy output. MIF is an attempt to avoid the physical and engineering problems of two opposing approaches to fusion, traditional magnetic confinement and inertial fusion.

As an impulse approach to thermonuclear power engineering, MIF has the main advantage over laser fusion – the presence of a magnetic field improves the confinement of particles and energy. The scheme of a laser fusion with a magnetic field works as follows. A pulsed solenoid creates an initial magnetic flux. After focusing the magnetic field, the liner is accelerated by a driver (laser, plasma or heavy ion). The field contracts and creates a dynamic pressure on the target shell from the inside, heating it up to ultrahigh temperatures. This increases the density of the magnetic flux, creating an ultra-strong magnetic field.

The proposed approach is an intermediate link between MCF and ICF, using high-density plasma and simultaneously small time scales, which is reflected in the following dependencies:

1) the reactivity of the plasma depends on the square of the density, which increases by several orders of magnitude in comparison with the traditional MTC;

2) all the characteristics of a plasma of length scale decrease together with the density;

3) the presence of a magnetic field in the plasma of the target significantly reduces the losses associated with the thermal conductivity in the liner material, which makes it possible to use relatively slow uniform compression and heating of the plasma to fusion conditions comparable to ICF, and weakens the requirements for the energy pulse to the values achievable by modern technology.

It is for these reasons that MIF was named a cheap way to implement the fusion. In this work we propose the use of a gaseous, or rather, a plasma liner, since such an approach has a low energy threshold for achieving economic profitability and the possibility of obtaining a high energy yield $Q > 5$.

Pulsed powerful sources of energy (drivers) are constructively simple and can be easily applied already at present taking into account the existing equipment and the current level of development of science and technology. Such an approach will potentially provide a cheap and fast research path to demonstrate practical fusion energy production and subsequent commercialization.

Traditionally, specially produced cold targets were used in the ICF for compression. At present, the technology is worked out and plasma targets of radius ~500 m are produced. The MIF assumes the use of 'hot' targets with a preheated plasma with a frozen field of ~1 T, a temperature of several electron volts (1 eV ≈ 11600 K) and a density of the order of 10^{17} m^{-3}, which can significantly reduce the requirements for the driver capacity. The main geometries and configurations of the targets are shown in Fig. 5.5.

The most promising magnetic configurations for MIF are the following systems.

Z-pinch (toroidal field only) is the simplest configuration, hard-core versions are possible for stabilization. The presence of a cylindrical shape assumes several equally probable compression scenarios, depending on the pinch stability. The plasma of the target is formed in the reactor chamber, it cannot form remotely, i.e., the entire load will be perceived by the walls of the reactor.

The FRC (poloidal field only) and spheromak (poloidal and toroidal field) belong to the class of compact torus (CT) installations – in experiments they show extraordinary stability, it is possible to form at a distance and then transport them to the reactor chamber, materials and lifetime of the reactor. The configurations can have a shape close to spherical to achieve uniform reduction of the target

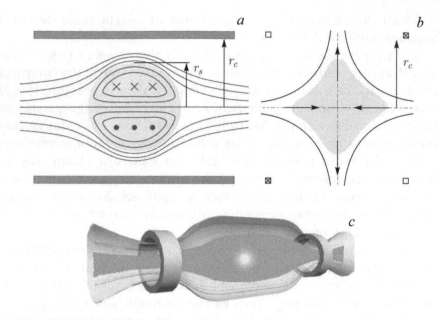

Fig. 5.5. Typical magnetic configurations for compression by lasers or supersonic jets: a – compact torus (FRC), b – cusp, $в$ – open trap.

without direct contact with the first wall. High parameters of the initial magnetized plasma are generated.

Also worth noting is the axisymmetric magnetic trap (cusp), which is easy to manufacture and experiments with which are being conducted at the present time. One of the varieties of the open magnetic configuration – the gas-dynamic trap – is considered in the introduction.

The compression striker (liner) can be a solid, liquid, gaseous or a combination of aggregate states. Its main purpose is to transfer its kinetic energy to the target and to work on compression. When using solid strikers because of the high density of materials, the liner transfers about 50% of its original energy, but this approach has a number of drawbacks. In order to circumvent them, a plasma liner is used, which in theory gives the same order of energy release. Next, we will refer to all types of liners, other than the plasma liner, the traditional liner (metal, conductive, exploding, multiwire, double liners, heavy ion and Z-pinch, extra-thin-walled and low-density liners, liner theta pinch, collapsible liners, etc.). Advantages of a plasma liner are:

1) the driver can be at a sufficiently remote distance from the target, which is difficult to implement in the case of traditional liners;

2) the cost of manufacturing metal liners limits economic profitability, which in turn increases initial capital investments;

3) solid fragments from the liner may pose a problem for the first wall of the reactor.

Next, consider the scheme of a pulse system of magnetically inertial fusion with a plasma liner (PULsed System and Alternative Reactor – PULSAR), which is an alternative installation with high-speed plasma guns (Fig. 5.6). Two preformed compact torus formations (spheromak or FRC) are injected into the combustion chamber by means of injectors, where plasma configurations collide in the centre and form a magnetized plasma. Plasma guns (plasma jet generators) are distributed in the chamber in such a way that the jets almost uniformly compress the plasma target and, merging, form a spherical or cylindrical plasma liner.

At MIF installations with plasma guns and jets, compression is accompanied by the formation of a powerful shock wave that converges to the axis. As a result, thermonuclear combustion will spread to the entire mixture contained within the shell. Separately note that, although only D–T-fuel is currently being considered, MIF allows us to move to lower temperatures (in comparison with the magnetic fusion): <10 keV for the D–T target, <100 keV for the $p–^{11}$ fuel B and <50 keV for the D–^{3}He-reaction.

With a dense compression, the plasma liner sends shock waves through the central plasma and shock heating occurs to elevated temperatures (~100 eV). High temperatures directly increase the

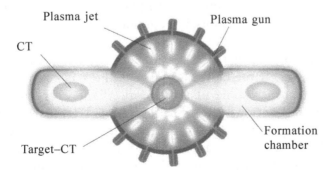

Fig. 5.6. The target and plasma liner formation scheme (above) and the conceptual diagram of the MIF reactor with plasma jets (bottom).

electric conductivity of the plasma to such a extent that the magnetic flux is retained inside the central plasma. Because of the prevailing radial pulse density of the plasma liner, it continues to compress the hot core toward the centre.

With further reduction of the target by a plasma liner and practically adiabatic heating to the conditions of thermonuclear burning, the magnetic flux is compressed together with the target, increasing the magnetic fields to a mega-gauss level. The compression is considered almost adiabatic due to the presence of a mega-Gaussian magnetic field that suppresses electronic and thermal conductivity, reduces losses by several orders of magnitude; the plasma density in the centre remains relatively low, so the losses due to bremsstrahlung are small.

The losses to electron synchrotron radiation are assumed to be insignificant. The rates of temperature loss are low enough, therefore, using plasma jets at a speed of 0.1 to 1 m in a microsecond, relatively slow compression and heating occur.

Such plasma jets can be obtained by means of coaxial guns using the Lorentz force. The plasma liner consists of two layers: the inner layer carries the main fuel for fusion, and the outer packed bed consists of a heavy gas, for example argon or xenon, or from a plasma of the same composition as the target fuel. Such a composite structure can be formed in one of two ways: as part of the process of charging plasma guns or synchronized launch of two sets of spherically located accelerators refueled with the required gas. The radial displacement of the plasma liner is restrained by the burning core and propagating beyond the shell of the wave, compressing the liner, form a cool layer of high-density fuel on the inside.

As a result of the increase in volume due to the formation of rarefaction waves and destruction of the plasma liner, the confinement of the nuclear burning nucleus is complete. The confinement time is approximately equal to the sum of the standing wave travel time and the double time of displacement of the discharge wave through the cold fuel layer. If the cold fuel layer is thick enough and dense to lock up and absorb some of the energy of the charged particles from the fusion reactions, then the thermonuclear fusion wave of the fusion will spread to the cold fuel and this will increase the energy yield (the ratio of released thermonuclear energy to the driver energy perceived by the target). However, a high coefficient of thermonuclear energy gain for this approach is not required, since the efficiency of plasma accelerators is quite large compared with lasers.

Understanding the processes of formation and existence of magnetic configurations gives us the entire spectrum of the dependences of the calculated plasma characteristics on external parameters which we can directly affect. To determine the equilibrium configuration, the problem is to find the solution of the Grad–Shafranov equation for a cylindrical flow outside the separatrix of the FRC (an external problem for open lines of force of a magnetic field).

The most widely used equilibrium – in the form of Solov'ev [335] – is an analogy of a Hill vortex [132]. Its advantage is that it is mathematically simple and really two-dimensional, but this model has a big minus. It does not have an analytical external solution. This problem represents a very serious defect, since the ion trajectories will extend into the region behind the separatrix.

However, there is an analytic solution for the field in the region outside the ellipsoid, which was obtained in [336]. If we interpret the flow function as a function of the magnetic flux ψ, then we have a magnetic field free of current around the elongated ellipsoid. We insert Solov'ev's solution inside and get a complete analytical solution for the field. It is not self-consistent, since even if ψ is continuous on the separatrix, then the magnetic field $B \sim$ grad ψ is not. It is necessary to sew the values of the field on the separatrix, but there is a jump in B, especially near the boundary, so problems in numerical simulation may arise, since ions need to cross regions where the magnetic field changes abruptly. The solution joining conditions are written in the same way as in [337]. The function of the internal magnetic flux is written in the form of Solov'ev, and the external solution written in the elliptic coordinate system (μ, ζ, ω) has the form

$$\psi_{\text{ext}} = \frac{1}{2} Ak(1-\mu^2)(\zeta^2-1)\left\{\frac{1}{2}\ln\frac{\zeta+1}{\zeta-1} - \frac{\zeta}{\zeta^2-1}\right\} +$$
$$+ CUk(1-\mu^2)^{0,5}(\zeta^2-1)^{0,5},$$

$$A = \frac{U_a}{\dfrac{1}{1-e^2} - \dfrac{1}{2e}\ln\dfrac{1+e}{1-e}}, \quad C = -\frac{1}{2}k(1-\mu^2)^{0,5}(\zeta^2-1)^{0,5},$$

$$k = ae, \quad \zeta_0 = 1/e,$$

where a, b are the polar and equatorial axes, and e is the eccentricity of the meridional section, U is the variable parameter.

Just like Solov'ev model, this approximation is an analytical number, which gives reason to hope that the solutions obtained can be used to form simple programs for simulating the trajectories of α-particles can be formed, and the contribution of their energy to the target.

To determine the equilibrium configuration, the problem is finding the solution of the Grad–Shafranov equation for a cylindrical flow outside the separatrix of the FRC. It is especially difficult to find solutions for the racer (the form of the separatrix 'stadium') of the configuration of the magnetic field, which is more often encountered in experiments.

Compression of a compact torus (FRC) is a possible way to MIF, and accordingly also to ignition of the fuel mixture in the combustion chamber. To achieve success in such experiments, a multi-parameter study is needed, but along this path there are numerous difficulties in modeling such processes. At the same time, the very presence of a laser or plasma driver leads to a complication and a rise in the cost of the system, but nevertheless it can work at much lower costs and modest parameters than the inertial fusion or magnetic fusion.

Physical processes include disparate time scales. For example, FRC has buffer regions of the field (near the vacuum field), in which the Alfven speed is high, while the implosion of the liner runs at much lower speeds. Such different time scales impose strict limitations and stringent requirements.

The success of the experiment depends on the real slowing of heat transfer by means of thermal conductivity to prevent energy losses from the plasma. Previous calculations of the final FRC compression stage very often produced a large error due to the inaccuracy of the numerical experiment when considering the parallel flow.

We have considered the listed shortcomings and made progress on these issues. A self-consistent model has been prepared for the axisymmetric configuration of a compact torus and the creation of a computer simulation of the processes of the final stages of target compression by plasma guns.

The proposed compression scheme assumes the use of formed targets based on various magnetic configurations. FRC is supposed to be used as a magnetized target. The dense plasma is retained by the initial magnetic field, then it is compressed by the liner and acquires high temperature and density. The parameters of the thermonuclear plasma obtained in this way should be such that the generated energy exceeds several times the initial kinetic energy of the liner. This requires adequate confinement of the compressed high-temperature plasma (which depends on the magnetic geometry), as well as the appropriate time duration of the strong magnetic field (which is determined by the dynamics of the liner's motion and the diffusion of the magnetic field into the liner material).

Depending on the method of creating the initial plasma, different magnetic configurations are obtained. Theoretical study allows us to determine the distribution of parameters before compression, which serve as initial conditions for further calculations of the various stages of reduction. Figures 5.7–5.9 show the results of 2D numerical simulation of various FRC configurations. R_s and Z_s are the geometric parameters of the plasma formation, the radius and the length of the separatrix, respectively. For further consideration, a spherical magnetic configuration with a radius $R_s = 0.10$ m, which is the initial (maximum) target radius, was selected.

When the target is compressed by plasma jets, it is necessary to take into account various phenomena occurring in the plasma core-liner system. Next, we consider the stage of maximum compression of the outer boundary of the magnetized plasma. Under the action of a powerful plasma jet on the target plasma, that is, directly at the moment of contact with the plasma liner, it is necessary to take into account the following phenomena: convective transport, radiation, ohmic heating, thermal conductivity and energy exchange with reaction products.

The interaction of fast plasma liners and a target or a rapid adiabatic compression has also disadvantages. Thus, one of the principal factors limiting the magnitude of the maximum achievable field in this case is the Rayleigh–Taylor instability. In this case, the external magnetic field suppresses other types of instabilities, for example, MHD and rotational modes. The consideration of instabilities is not included in the range of questions presented in this chapter; therefore, we restrict ourselves to a reference to papers in this field [338, 339].

Accurate calculations by inertial compression are based on numerical experiments that take into account the interaction of various physical processes occurring in the plasma. However, the simplest model allows us to evaluate some of the principal stages that occur in the scheme under consideration, namely, the formation of a hot core, the maintenance and propagation of combustion, i.e., the emergence of a self-sustaining thermonuclear reaction or the physical threshold of a fusion reaction (breakeven). In this case,

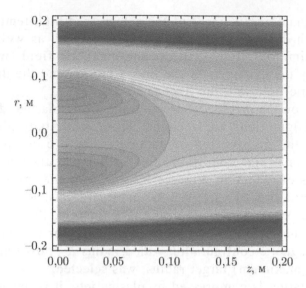

Fig. 5.7. The magnetic flux function ψ for the compact torus FRC in the (r, z) coordinates at $R_s = Z_s = 0.10$ m, $B = 10$ T. Sphere configuration.

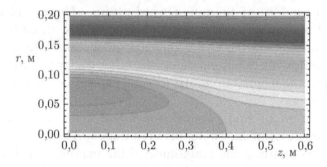

Fig. 5.8. The magnetic flux function ψ for a compact torus FRC in coordinates (r, z) at $R_s = 0.1$ m, $Z_s = 0.3$ m, $B = 10$ T. Configuration of the ellipse

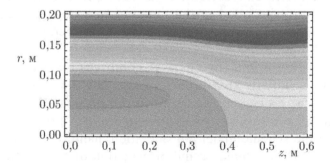

Fig. 5.9. The magnetic flux function ψ for a compact torus FRC in coordinates (r, z) at $R_s = 0.1$ m, $Z_s = 0.3$ m, $B = 10$ T. Stadium configuration.

the energy balance is considered for a sphere of uniformly heated fuel surrounded by a shell of a colder substance (liner). The rate of change in the internal energy density E for a hot nucleus is written in the form

$$\frac{dE}{dt} = P_{dep} - P_m - P_{rad} - P_{cond},\qquad(5.22)$$

where p_{dep} is the power invested in the target by the products of the reaction, p_m is the specific loss due to mechanical work, p_{rad} and p_{cond} are the powers of radiation losses and due to thermal conductivity.

The balance equation is written for a homogeneous sphere of equimolar D–T-fuel having a radius R, density ρ and temperature T, which is the same for ions and electrons. The specific quantities occurring in the right-hand side of Eq. (5.22) are averaged over the volume under consideration occupied by the plasma.

The deuterium–tritium (D–T) reaction is considered:

$$D + T \to \alpha(3.52\ \text{MeV}) + n(14.07\ \text{MeV}).\qquad(5.23)$$

The energy obtained as a result of the reaction can be written down as

$$p_{fus} = 5p_\alpha = 5A_\alpha \rho^2 \langle \sigma v \rangle,\qquad(5.24)$$

where p_α is the energy contribution of α-particles, $A_\alpha = 8 \times 10^{39}$ J/kg^2, $\langle \sigma v \rangle$ is the rate of the thermonuclear reaction averaged over the Maxwellian velocity distribution spectrum. It is possible to write down the contribution of the energy of the fusion products to the target

$$p_{dep} = p_{fus} f_{dep} = p_\alpha (f_\alpha + 4f_n),\qquad(5.25)$$

where f_α and f_n are the fractions of the α-particles and neutrons.

Next, let us consider in more detail the interaction of reaction products fusion with background plasma. Approximate expressions are used for quantitative estimates of processes occurring in a D–T plasma at temperatures up to 20 keV, typical of MIF. In the absence of a magnetic field in the target, the largest retardation of α-particles will occur in collisions with electrons at a small angle. Alpha particles move almost along a straight trajectory, and their velocity decreases according to

$$\frac{dv_\alpha}{dt} = -\frac{v_\alpha}{2\tau_{ae}},$$ (5.26)

where $\tau_{ae} \approx 42 T_e^{3/2} / \rho \ln \Lambda_{ae}$ is the time for the energy contribution, $\ln \Lambda_{ae}$ is the Coulomb logarithm characterizing collisions between α-particles and electrons, and T_e is the electron temperature (in keV).

The transfer of thermonuclear energy to the reacting central plasma occurs in two stages: α-particles are heated by electrons, which in turn interact with ions during the time τ_{ei}. It turns out that $\tau_{ae} = \tau_{ei}$. Most complex compression models with laser beams allow one to take into account different temperatures for the electron and ion components, as well as the finite retardation time of the α-particle. In our estimation calculations with a plasma liner, it is assumed that the α-particle gives its energy to the plasma instantaneously due to elastic collisions. The electrons and ions of fuel have the same temperature. This approximation is adequate enough if the times τ_{ae} and τ_{ei} are much shorter than the time required for a self-sustaining reaction to occur.

The distance that an α-particle with an energy of 3.52 MeV can overcome in a homogeneous plasma can be obtained as $l_\alpha = \int_0^x v_\alpha dt$, using the expression (5.26) given above, we obtain

$$l_\alpha = 2v_{\alpha 0}\tau_{ae} = 0.107 \frac{T_e^{3/2}}{\rho \ln \Lambda_{ae}},$$ (5.27)

where $v_{\alpha 0}$ is the initial velocity of the α-particle. The quantity l_α does not depend on the degree of compression and is determined only by the temperature T and the density ρ. At typical plasma parameters of MIF $\rho \approx 100$ kg/m^3, $T \approx 100$ eV, the mean free path of the fusion products is many times greater than the target radius. And the product of the density by the mean free path for the D–T mixture is much larger than the optical thickness of the plasma ρR. Thus, a strong magnetic field is a means of retaining high-energy particles. Let's estimate its value:

$$v_\alpha = \left(\frac{2E_{\alpha 0}}{m_\alpha}\right)^{1/2} = 1.4 \cdot 10^7 \, \text{m/s}.$$ (5.28)

Then the initial radius of the synchrotron circle is

$$r_c = \frac{v_\alpha m_\alpha}{eB}. \qquad (5.29)$$

To effectively heat a plasma with charged particles, it is required that this value does not exceed the characteristic size of the liner cavity in the final state. Typical sizes of configurations used for compression are ~0.1 m. Spherical convergent shock waves can compress a target with a factor of 33 [340], but the presence of instabilities reduces the compression ratio. Taking into account the actual processes involved in the reduction of the target, the compression ratio in this work is assumed to be 20, and the target radius of the target is approximately 0.005 m. To keep the α particle from the formula (5.29), induction of a magnetic field of not less than $B \approx 100$ T is required. We note that, depending on the design of the plasma guns, the method of dispersing the liner, the formation of plasma jets, their focusing and uniform reduction, in the experiment a magnetic field of 500 to 1000 T.

It should be noted that the α-particles leaving the hot core are very effectively stopped by an outer shell consisting of colder and often denser fuels. Thus, the transfer of the α-particles leads to rapid heating of a thin layer of matter located outside the hot core, this process promotes the propagation of the combustion wave of the fuel.

The direct heating of the plasma of the D–T reaction depends on the ratio of the deceleration time to the confinement time of the α-particle. To determine the approximate dependence of the contribution of the energy of α-particles to the target from the magnetic field, the results of numerical simulation [341] were generalized for the cases of a static azimuthal field having two strongly different gradients and a homogeneous azimuthal field:

$$f_\alpha = \frac{\rho(C(BR)^2 + R)}{\rho(C(BR)^2 + R) + D}, \qquad (5.30)$$

where ρ is the density of the target plasma, BR is the parameter determined at the plasma boundary in the median plane, C and D are constants. The constant C depends on the plasma temperature, the magnetic field gradient, and the target radius. The constant D characterizes the adsorption of α-particles in the target in a zero magnetic field, then $D = 2$ kg/m^2 is used. In Fig. 5.10, where the proportions of α-particles contributing to the fuel heating are shown, depending on the BR parameter, the following values are typical for MIF: *1 – C* = 1 T^{-2}·m^{-1} at $\rho R = 0.5$ kg/m^2, *2 – C* = 0.117 T^{-2}·m^{-1} for $\rho R = 0.1$ kg/m^2, *3 – C* = 0.01 T^{-2}·m^{-1} for $\rho R = 0.01$ kg/m^2.

The dependence of the energy contribution of α-particles to the target from parameters of plasma is shown in Fig. 5.11.

Neutrons with an energy of 14.07 MeV, produced as a result of D–T-reactions, interact with the plasma core, mainly through elastic collisions. On average, they lose $2A/(A + 1)^2$ part of their energy when they collide with nuclei with mass number A. The contribution of neutron energy to the hot nucleus can be neglected because of the smallness of magnitude. However, it should be taken into account when exact models of fuel combustion are studied or a large volume of flammable fuel is set. The contribution of neutrons to the target plasma is defined as

$$f_n = \frac{\rho R}{\rho R + H_n},$$ (5.31)

where H_n = 200 kg/m² is a constant that corresponds to a homogeneous D–T sphere with a constant source of neutrons. Figure 5.12 shows the fraction of neutrons as a function of temperature and the optical thickness of the D–T target as a result of compression by converging plasma jets of a compact torus.

Bremsstrahlung losses are x-rays, which can not be directly converted to electricity, as is possible in the case of charged particle energy or synchrotron radiation. Figure 5.13 illustrates the radiation losses as a function of the radius of the compressible magnetized target. The temperature and concentration increase with

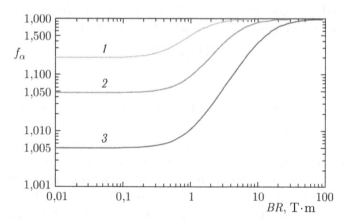

Fig. 5.10. The fraction of the contribution of the energy of α-particles (charged products of the D–T reaction) to the target plasma from the parameter BR: *1* – $\rho R = 0.5$ kg/m²; *2* – $\rho R = = 0.1$ kg/m²; *3* – $\rho R = 0.01$ kg/m².

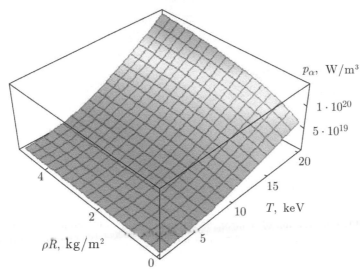

Fig. 5.11. The specific power of the contribution of α-particles to the target plasma when the D–T mixture burns as a result of compression at $B = 100$ T. Hereinafter, T is the average temperature of the compressed fuel and ρR is the optical thickness of the compressed fuel

the contraction of the target, as shown in [342], according to $T = T_0(V_0/V)^{\gamma-1}$ and $n = n_0(V_0/V)$, where T_0, n_0 and V_0 are the initial temperature, concentration and the volume of the plasma configuration (target). In the special case of the absence of energy losses from the plasma, the pressures of the gas and magnetic components vary according to the adiabatic law with the adiabatic exponents: for the gas $\gamma = 5/3$ and the magnetic field $\gamma = 4/3$ (the case of isotropic spherical compression) [343]. The power of bremsstrahlung is shown in Fig. 5.14.

Losses on synchrotron radiation are negligible compared to the rest, especially for systems with high beta, which is typical for compact systems such as a torus or a mirror. Therefore, this kind of energy is taken equal to 0, i.e. $p_{\text{rad}} = p_{\text{br}}$.

The target exchanges energy with the surrounding plasma by means of a mechanical work [330], which for a homogeneous sphere can be written (the graph is shown in Fig. 5.15):

$$p_m = \frac{1}{V}\frac{dE}{dt} = \frac{p}{V}\frac{dV}{dt} = p\frac{S}{V}u, \qquad (5.32)$$

where u is the surface velocity of the sphere, p is the pressure, $S/V = 3/R$ is the ratio of the surface area to the volume of the liner

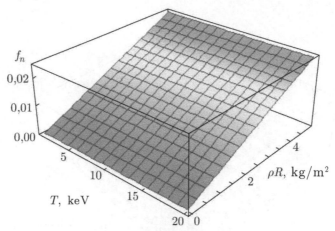

Fig. 5.12. The fraction of neutrons in the target MIF in the combustion of D–T-fuel due to a compressed spherical plasma liner.

cavity. Using the equation of state of an ideal gas $p = \rho R_{DT}T$, where $R_{DT} = 7.66 \cdot 10^{10}$ J/(kg·keV) is the gas constant for the D–T reaction, we obtain

$$p_m = 3\frac{pu}{R} = 3\frac{R_{DT}\rho Tu}{R}. \tag{5.33}$$

The loss power, written in the form of a volume flow of energy, in the quasispherical approximation

$$P_{\text{cond}} = \chi\frac{T}{3}\left(\frac{S}{V}\right)^2, \tag{5.34}$$

where χ is the Braginskii transport coefficient [344]. The power of heat losses during compression by a high-speed liner is shown in Fig. 5.16. The compression of a spherical target is assumed to be uniform with simultaneous convergence of plasma jets.

5.4. Modes and limiting amplification in the target MIF

Consider a scheme with a direct compression of the magnetized target. The source of energy are laser beams (laser driver) or plasma guns (plasma liner).

Omitting the material science and technology problems of thermonuclear fusion, let us turn to the physics of heating the target, or more precisely, to the estimates of the target confinement time

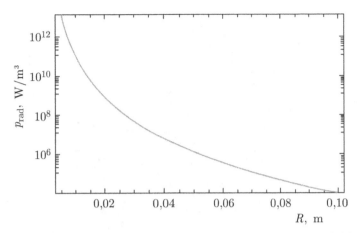

Fig 5 13 Radiation losses as a function of target radius *R*. The initial target parameters $T_0 = 2$ eV, $n_0 = 10^{21}$ m^{-3}.

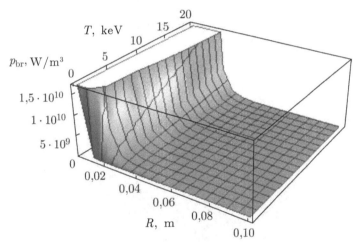

Fig. 5.14. The bremsstrahlung capacity of the compressible plasma FRC as a function of the target parameters (R is the radius of the magnetized plasma, T is the temperature).

necessary to transfer energy from the external source (driver) to the magnetized plasma.

To realize a self-sustaining thermonuclear reaction (ignition), the following condition must be fulfilled: the product of the plasma density, its temperature and the energy confinement time must exceed a certain value. The product of the density of the temperature is proportional to the pressure, which can be expressed in terms of the magnitude of the magnetic field. For an effective magnetic field B we

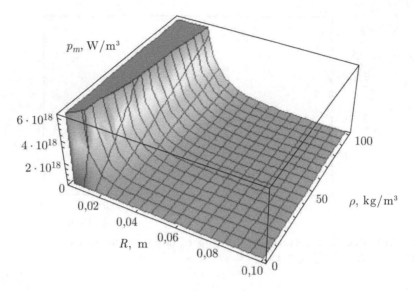

Fig. 5.15. Mechanical target losses at a compression rate $u = = 1.25 \cdot 10^5$ m/s; R is the radius, ρ is the plasma density.

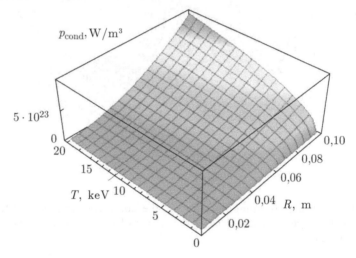

Fig. 5.16. Specific heat losses of magnetized plasma due to thermal conductivity.

take the value corresponding to the equality of the plasma pressure p and the magnetic pressure

$$p_M = \frac{B^2}{2\mu_0},$$ (5.35)

where $B^2/2\mu_0$ is the magnetic field pressure (magnetic pressure), μ_0 is the magnetic constant.

Thus, the ignition condition reduces to a certain value of the product $B^2\tau$, where τ is the energy confinement time determined by the processes of heat conduction and convection. In particular, for a deuterium–tritium plasma with a temperature of 10 to 20 keV, the ignition is realized at $B^2\tau \approx 3$ T²·s.

Let us assume that the convective energy loss per ion (electron) is $\frac{3}{2}k_B T_i \left(\frac{3}{2}k_B T_e\right)$, where k_B is the Boltzmann constant, T_i is the ion temperature, and T_e is the electron temperature. Then the confinement time of the internal energy of the plasma is determined by the particle confinement time. In the system under consideration, the convective heat transfer with respect to the direction of the magnetic field lines can be divided into longitudinal and transverse components. The contribution of the classical transverse thermal conductivity in the conditions under consideration is negligibly small.

The longitudinal flow has a practically gas-dynamic character, since the plasma density is large, and the area of the 'slits' through which the plasma flows out is small. The longitudinal flow rate is of the order of the thermal velocity of the ions: $v_{T_i} = \sqrt{k_B T_i / m_i}$ (m_i is the average ion mass). The total flux of particles along the magnetic field lines is

$$\left(\frac{dN}{dt}\right)_{\parallel} = (n_i + n_e)v_{T_i}(S_0 + S_L), \tag{5.36}$$

where N is the total number of particles in the plasma sphere, S_0 is the area of the axial (point) cusp, S_L is the area of the annular (sometimes called linear) cusp gap.

The size of the exit regions of particles from the cusp is as follows: the radius of the axial gap is $\delta_0 \approx 2\rho_i$, the width of the annular gap is $\delta_L \approx 2\delta_0^2 r_c$, where ρ_i is the average thermal synchrotron radius of the ion (calculated from the effective magnetic field), r_c is the radius of the cusp. We take the following relationship between the radii of the plasma sphere and the cusp: $a \approx 0.6r_c$. The areas are equal to $S_0 = 2\pi\delta_0^2 \approx 8\pi\rho_i^2$ and $S_L = 2\pi r_c \delta_L \approx 4\pi r_c \delta_0^2 r_c \approx 16\pi r_c \rho_i^2 r_c$. The total number of particles is $N = \frac{4}{3}\pi a^3(n_i + n_e)$. The time of longitudinal losses

$$\tau_{\parallel} = N / \left(\frac{dN}{dt}\right)_{\parallel} = \frac{2a^3}{3vT_i(\delta_0^2 + r_c\delta_L)}. \tag{5.37}$$

Similarly, the time $\tau_\perp$ of losses is introduced across the magnetic field lines. This time can be taken equal to the Bohm time corresponding to the maximum intensity of turbulent plasma transport across the magnetic field. Then

$$\tau_\perp = N / \left(\frac{dN}{dt}\right)_\perp = \frac{q_i B a^2}{k_B T_i},$$ (5.38)

where q_i is the average charge of the ion, the numerical factor in the Bohm dependence is ignored, which corresponds to the most pessimistic mode of transverse losses.

The resulting loss time

$$\tau = (1/\tau_\parallel + 1/\tau_\perp)^{-1}.$$ (5.39)

Since the most stringent requirements are imposed on the parameters of the plasma at the final stage of the compression, we present the results of the estimates for this particular case. For calculations, we take $T_i = T_e$. The results of calculations of the effective field and the corresponding values of the confinement time are shown in Figs. 5.17 and 5.18. Two limiting cases were considered. In the first case (line 1 in Figs. 5.17, 5.18) both longitudinal and transverse losses are taken into account, in the second (line 2), the transverse losses are neglected. Obviously, the second case is preferable, but the question of how turbulent transverse losses can be suppressed, of course, requires detailed investigation. Thus, the region I in Fig. 5.17, 5.18, limited by these limits, corresponds to the parameters of magneto-inertial (MI) confinement of plasma.

We note that the effective field values in the MI regime are extremely high, but theoretically achievable under impulse conditions. Further decrease of the required field value is possible when taking into account the ambipolar electrostatic potential of the plasma, which, apparently, will be formed to maintain the quasineutrality of the plasma. This regime can be called magnetic-inertial-electrostatic (MIE) confinement. In Figs. 5.17, 5.18 the MIE-regime corresponds to region II.

The power of bremsstrahlung depends on the temperature and electron density and is independent of the shape and dimensions of the plasma formation, in contrast to the synchrotron losses, which are also determined by the values of the magnetic field, which is essential for the MI regime. But taking into account the current level of development of science and technology, it is considered that

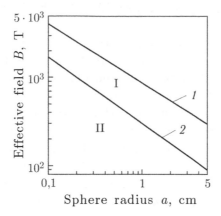

Fig. 5.17. Modes of convective heat transfer of a hot quasispherical plasma in a magnetic field of counter ring currents at the final stage of compression: I – magnetically inertial, II – magnetically inertial–electrostatic.

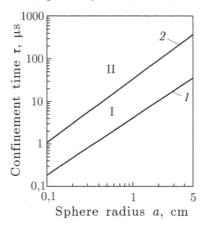

Fig. 5.18. The confinement times corresponding to the regimes shown in Fig. 5.17.

synchrotron radiation is almost completely returned to the plasma, since the average absorption of radiation by the plasma at multiple reflections from the walls is ~0.9. Or, in other words, the coefficient of reflection from the walls is ~0.9, that is, the synchrotron loss power can be neglected.

Speaking of losses, it must be borne in mind that adiabatic invariance is violated near the centre of the cusp and the plasma from this region is rapidly lost along the magnetic field lines. To eliminate these losses, it is possible to use a system of special electrodes in the axial and annular gaps that prevent the loss of electrons. Then the ion confinement will be provided by the intrinsic ambipolar plasma potential.

Together with the plasma particles, the magnetic field will also confine charged reaction products in the trap. The study of confinement of fast alpha particles and plasma heating is an important issue that is beyond the scope of this chapter.

As a variant of the target for MIF with compression of plasma jets, one can consider a plasma of a cylindrical or spherical shape surrounded by a magnetic field. If the plasma pressure is high, then the magnetic field forms a kind of internal magnetic shell (IMS, afterburner, that is, the inner layer of the liner) outside the plasma. Thus, inside the IMS there is a target, outside – a plasma liner formed by converging jets. The magnetic field is compressed under the action of the liner and compresses the target. Ideally, during compression, the pressure has practically the same value for the target plasma, magnetic field and liner plasma. The diffusion of the magnetic field in the plasma of the target and the liner can be neglected, since the compression process in the MIF modes occurs much faster.

An important function of IMS is the thermal insulation of the target plasma from the liner. This makes it possible to obtain at the final stage the temperature of the target T, several orders of magnitude higher than the temperature of the liner T_L. In the first approximation, we can assume that, after compression, the target expansion is determined by the thermal velocities of the liner plasma, and therefore the plasma confinement time in the MIF scheme under consideration will be several orders of magnitude higher than the confinement time for ordinary inertial fusion. Consequently, the question of the realization of regimes with $T_L \ll T$ in a target with IMS is fundamentally important from the point of view of the prospects of such systems. The limiting factor, apparently, is the magnitude of the magnetic field.

Let us consider the limiting efficiency that can be achieved in the considered target in the framework of the simplest approximations. The efficiency of the target–IMS–liner scheme can be characterized by the plasma gain

$$Q_{pl} = \frac{W_{fus}}{W_{pl} + W_M + W_L}, \qquad (5.40)$$

where W_{fus} is the fusion energy released during the confinement, W_{pl}

is the thermal energy of the plasma, W_M is the energy of the magnetic field, W_L is the energy of the liner.

The results of the evaluations are presented in Table 22. The compression of a target consisting of equal parts of deuterium and tritium, which have a temperature T_0 before compression, is considered. The target scheme is shown in Fig. 5.19. The target of radius r_0 is placed in a magnetic field with induction B_0, occupying a region of radius R_0. As a result of the action of plasma jets around the target, a plasma liner is created with a temperature T_L compressing the magnetic field it covers before induction B.

In this case, the outer radius of the IMS decreases to R, the radius of the target decreases to r, and its temperature increases to the thermonuclear value $T \approx 10-15$ keV. Compression is considered adiabatic. For a target, one can consider both a cylindrical regime (a long target) and a quasispherical regime. At identical compression ratios r/r_0, the second mode allows for more heating, so we considered quasi-spherical compression of the target with an adiabatic exponent of 5/3. The compression of the magnetic field is two-dimensional, and it was considered with allowance for the conservation of the magnetic flux. The confinement time was estimated as follows: $\tau = r(k_B T_L/m_L)^{-1}$, where k_B is the Boltzmann constant, m_L is the mass of the liner ions. In addition to the energy W_{fus} released during the reaction during the confinement time, Table 2.2 also gives the energy W_{fus0}, which could be released with complete combustion of the fuel.

The target material (thermonuclear fuel) consists of equal parts of deuterium and tritium. The liner material is an easily ionized gas with a relatively high atomic mass, in the case under consideration – argon or xenon.

5.5. Source of neutrons based on MIF

Below we analyze the possible application of magneto-inertial fusion (MIF) to create a neutron source. The MIF approach involves heating the magnetized plasma formation to thermonuclear temperatures by compression. As a result of compression, not only high temperatures (~10 keV) are reached, but also high densities (~10^{27} m^{-3}). The required plasma confinement time corresponds to the time of inertial expansion. Table 2.3 shows the parameters of laser installations and Z-pinch, which can be used as MIF drivers.

Table 22. Parameters of target with IMS and plasma liner

	Var. 1	Var. 2	Var. 3
Parameters before compression			
Target radius r_0, cm	10	10	10
The outer radius of the IMS R_0, cm	17	17	17
The magnetic field B_0, T	5,5	3,8	5,7
The density of deuterium n_{D_0}, m^{-3}	$7.5\cdot10^{23}$	$2.5\cdot10^{23}$	$5.3\cdot10^{23}$
Target temperature T_0, eV	25	37,5	37,5
The substance of the liner	Ar	Ar	Xe
The liner temperature T_L, eV	10	1	10
The density of the energy flux of the liner to the target J, W/m^2	$5.7\cdot10^{18}$	$9.6\cdot10^{17}$	$3.5\cdot10^{18}$
Parameters after compression			
Target radius r, cm	0.5	0.5	0.5
External radius of the IMS R, cm	0.6	0.6	0.6
Liner radius L, cm	1.2	1.2	1.2
Magnetic field B, T	9824	6950	10155
Deuterium concentration n_D, m^{-3}	$6\cdot10^{27}$	$2\cdot10^{27}$	$4,3\cdot10^{27}$
Target temperature T, keV	10	15	15
Plasma energy (thermal) W_{pl}, J	$2.3\cdot10^7$	$1.1\cdot10^7$	$2.4\cdot10^7$
The energy of the magnetic field W_M, J	$6.6\cdot10^6$	$3.3\cdot10^6$	$7.1\cdot10^6$
The energy of the liner W_L, J	$5.9\cdot10^7$	$2.9\cdot10^7$	$6.2\cdot10^7$
Parameters of the burn phase			
Fusion energy W_{fus}, J	$1.5\cdot10^8$	$1.2\cdot10^8$	$3.2\cdot10^8$
Fusion energy potential W_{fus0}, J	$6.6\cdot10^9$	$2.2\cdot10^9$	$4.7\cdot10^9$
The plasma confinement time τ, s	$3\cdot10^{-8}$	$1\cdot10^{-7}$	$5.8\cdot10^{-8}$
Plasma gain Q_{pl}	1.7	2.8	3.5

Table 22 (continued)

	Var. 4	Var. 5	Var. 6
Parameters before compression			
The substance of the liner	Ar	Xe	Xe
The magnetic field B_0, T	6,3	9,0	9,0
The liner temperature T_L, eV	1	10	1
The density of the energy flux of the liner to the target J, W/m^2	$2.5 \cdot 10^{18}$	$8.8 \cdot 10^{18}$	$2.8 \cdot 10^{18}$
Parameters after compression			
Magnetic field B, T	11340	16040	16040
Deuterium density n_D, m^{-3}	$4.0 \cdot 10^{27}$	$8.0 \cdot 10^{27}$	$8.0 \cdot 10^{27}$
Target temperature T, keV	20	20	20
Plasma energy (thermal) W_{pl}, J	$3.0 \cdot 10^7$	$6.0 \cdot 10^7$	$6.0 \cdot 10^7$
The energy of the magnetic field W_M, J	$8.8 \cdot 10^6$	$1.8 \cdot 10^7$	$1.8 \cdot 10^7$
The energy of the liner W_L, J	$7.7 \cdot 10^7$	$1.5 \cdot 10^8$	$1.5 \cdot 10^8$
Parameters of the burn phase			
Fusion energy W_{fus}, J	$7.8 \cdot 10^8$	$1.8 \cdot 10^9$	$5.6 \cdot 10^9$
Fusion energy potential W_{fus0}, J	$4.4 \cdot 10^9$	$8.8 \cdot 10^9$	$8.8 \cdot 10^9$
The plasma confinement time τ, s	$1.0 \cdot 10^{-7}$	$5.8 \cdot 10^{-8}$	$1.8 \cdot 10^{-7}$
Plasma gain Q_{pl}	6.7	7.7	24

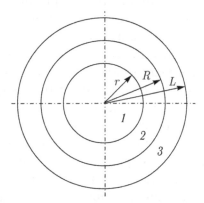

Fig. 5.19. The structure of a target with an internal magnetic shell (IMS) and a liner (section in the plane perpendicular to the magnetic lines of force): *1* – target plasma, *2* – IMS, *3* – plasma liner.

High gain corresponds to lasers, ion beams (as drivers), for example, installations of OMEGA EP, Z-Beamlet laser. For typical MIF scenarios, 10 000 T magnetic fields are required, which are realized as a result of compression of the initial magnetic flux. Ignition in this case is possible at a low implosion rate due to the magnetization of the target.

Medium and low power amplification factors provide plasma jets and various airliners (Shiva Star-FRX-L, CTIX, Tor-Liner). The gain factor for economically feasible energy production can be much lower than for the laser ICF driver. This also includes solid and liquid shells (liners), non-cryogenic gas targets and high-efficiency cheap drivers. For example, the current accelerator HyperV Plasma Jet and installation of the Plasma Liner physics exploratory eXperiment (PLX).

In this section, the configuration of the target in the form of an axially symmetric open trap with magnetic mirrors is considered. From the technical point of view, the essential advantage of the mirror is its simplicity. In addition, axial symmetry removes the problem of neoclassical transfer. For stationary and quasi-stationary open systems, the stabilization problems of the characteristic magnetohydrodynamic and kinetic instabilities are significant. The impulsive nature of magnetic–inertial regimes removes these problems. For the confinement times typical for MIF, magnetohydrodynamic instabilities do not have time to develop. Turbulent transport caused by kinetic instabilities has a significantly lower intensity compared with losses along magnetic lines of force. Thus, the efficiency of confinement of particles and energy in open traps is determined by the classical longitudinal losses, which limits the possibilities of stationary systems. At confinement times typical for MIF, the role of longitudinal losses in the energy balance of the plasma decreases, which is associated with a high plasma density and, accordingly, with high magnetic fields. In this paper, we consider such regimes in which the longitudinal loss time is comparable or exceeds the time of inertial expansion of the plasma.

The thermonuclear regimes of a quasistationary axially symmetric open trap, for which the longitudinal losses are determining, were considered in Refs [92, 93]. A feature of the regimes considered was the powerful injection of fast particles into the plasma. As shown by the simulation, modes with a power amplification factor in the plasma $Q_{pl} \approx 1$ (Q_{pl} is the ratio of the fusion power to the absorbed power of external heating) can be realized in a relatively compact system. Such

Table 23. Installations for generating high energy densities and for using as drivers

		Laser installations and generators					
	UFL-2M (RF-NC-VNIIEF)	NIF (LLNL, Livermore)	LMJ (CESTA, Bordeaux)	Gekko (ILE, Osaka)	OMEGA (University of Rochester)	Iskra-5 (VNIIEF, Sarov)	LFEX (ILE, Osaka)
Energy input parameters	4.6 MJ	3.6 eV on a particle	3.6 eV on a particle	10^{15} W	1 PW, 1 kJ per ps	10^{15} W, 100 fs, 100 kJ	2 ps, 5 PVT
Pulse duration		1–20 ns	9 ns	0,5–1 ps	1–100 ps	3–4 ns	1–20 ps
Spot size (wavelength)	1053 nm	0.3 mm	0.35 μm	20 μm (1.05 μm)	1053 nm	— (0.35 μm)	20–30 μm
Energy in impulse	2.8 MJ	1.8 MJ	2 MJ	500 J (beam)	2.5–60 kJ	600 kJ	10 kJ
Intensity (W/cm²)	527 nm – target irradiation	$2 \cdot 10^{15}$ indirect compression	10^{15} indirect compression	10^{20}	$2 \cdot 10^{20}$ direct compression	10^{18}–10^{21}	10 pp

Table 23. (End)

	Baikal (Troitsk, TRINITI)	Z-pinch					
		Z-Machine (Sandia National Laboratory)		S-300 (Kurchatov Institute)		Angara 5-1 (TRINITI)	
		Current	X-ray	Current	X-ray	Current	X-ray
Energy input parameters	50 MA	20 MA	50–250 eV	5–10 MA/cm	100 eV	1 MA/cm	100 eV
Pulse duration	150 ns	100 ns (front)	5–15 ns	100 ns	12 ns	90 ns	6 ns
Spot size	Inductively 900 MJ	—	1 mm (cylinder)	—	2 mm	—	2 mm
Energy in impulse	To the liner 50 MJ	16 MJ	1.8 MJ	300 kJ	50 kJ	600 kJ	120 kJ
Intensity (W/cm^2)	4–6 MV, 3 MA, 100 ns	—	10^{14}	—	$2 \cdot 10^{12}$	—(Power 12 TW)	$3 \cdot 10^{12}$

a system is comparable in size to existing experimental installations of open type, such as, for example, the gas-dynamic trap. Note that powerful injection in such systems contributes to the formation of positive potential barriers at the ends of the open system, which in turn improves longitudinal confinement. The principal difference between the magnetically inertial regimes considered here is the high density of the plasma and the correspondingly high energy density. Plasma heating in MIF mode is also carried out in a fundamentally different way – due to compression.

The main objective of this study is to analyze under what conditions in the MIF system, based on an open trap, the regimes with $Q_{pl} \sim 1$ can be realized, which is necessary to determine the direction of further development of the proposed concept.

An important advantage of open traps in comparison with classical tokamaks is the possibility of a stable plasma confinement by a high ratio of β plasma pressure to magnetic pressure. In particular, regimes with $\beta \approx 0.5$–0.6 are realized on the GDT.

To simulate the kinetics of fast ions, a physical model was developed [25, 112], taking into account the angular scattering of fast particles, as well as their participation in thermonuclear reactions. The MIF model is used for the analysis with compression of laser beams and plasma jets [257, 312]. One of the main indicators of efficiency is the power amplification factor in plasma given by equation (5.40). T

he results of calculations are presented in Table 24. Magnetic-inertial regimes are presented in the variants 1–3. For comparison, typical parameters of a quasi-stationary neutron source are also given [92, 93].

As can be seen, the power gain factor $Q_{pl} \approx 1$ in the MIF modes exceeds the typical values for a similar quasistationary system with magnetic confinement. This is due to the achievement of high plasma density as a result of its impulse compression. The source of thermonuclear neutrons with $Q_{pl} \sim 1$ looks attractive as a driver of a hybrid fusion reactor, and the relative simplicity of the considered target configuration and moderate requirements for parameters (with respect to typical inertial schemes) opens certain prospects for the development of this direction.

Table 24. Parameters of the magneto-inertial regimes of a cylindrical target (variants 1–3) and a quasi-stationary neutron source (QSNS) based on an open trap [92, 93]

Parameter	Var. 1	Var. 2	Var. 3	QSNS
Radius of plasma a, cm	0.25	0.5	0.25	100
Length of plasma L, cm	10	10	10	1000
The magnetic field of the central solenoid B_0, T	3300	1500	3300	1.5
A magnetic field in mirrors B_m, T	10000	15000	10000	11
Density of fuel $n_D = n_T$, m^{-3}	$1.3 \cdot 10^{27}$	$1.3 \cdot 10^{26}$	$6.4 \cdot 10^{26}$	$2.6 \cdot 10^{19}$
Plasma temperature T, keV	5	10	10	10
Plasma energy W_{pl}, MJ	12*	10*	12*	10,5
The magnetic field energy W_M, MJ	12*	10*	12*	—
W_L, MJ Liner energy	32*	27*	32*	–
Fusion power P_n, MW	43**	36**	63**	24
Power in neutrons P_n, MW	34**	29**	50**	24
Plasma gain Q_{pl}	0.77	0.77	1.1	0.38
The yield of neutrons N, 10^{18} neutrons/sec	15**	13**	22**	11

* In the pulse;
** Average value at a pulse repetition rate of 1 Hz.

6

Appendix

Below we present an analysis of the calcul0ated dependences for cross sections and rates of fusion reactions. Most modern computer simulations of fusion reactions use functions based on data that were published back in the 70–80s of the last century. Over the past thirty years, improved experimental methods have been developed, giving more accurate values, especially at low plasma temperatures. The effective cross section of the reaction and the velocity are

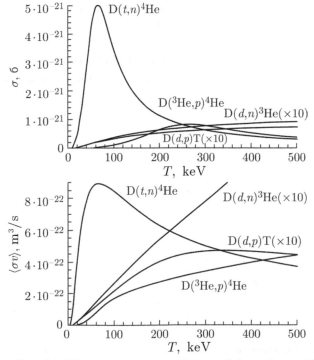

Fig. 6.1. The cross sections and rates of the main reactions.

calculated in the centre of the mass system, since the thermonuclear reaction proceeds over the entire volume at once, for the rates of fusion reactions the total energy is taken into account. The changes of σ even by a few percent can lead to significant changes in the parameters of future thermonuclear reactors. In turn, this affects the speed coefficient. The derived equations are the result of an analysis of the last experimental data of the effective cross section and the model of the fusion reactions. Figure 6.1 contains the cross sections and rates of the main reactions, which we will call the main ones, constructed for the range from 0 to 500 keV.

Figure 6.2 shows a comparison of the data [81] and [345, 346]. The two branches of the D–D reaction are combined together. Energy is stored in a logarithmic scale.

The approximate formula for the cross sections of the four main reactions in the interval required for the thermonuclear fusion has the form

$$\sigma = (A0 + A1 \cdot x + A2 \cdot x^2 + A3 \cdot x^3 + A4 \cdot x^4 + A5 \cdot x^5) \cdot 10^{-28}, \text{ m}^2.$$

The corresponding coefficients are given in Table 25.

The dependence of the velocity coefficient on the reaction temperature is shown in Fig. 6.3. The notation is the same as in Fig. 6.1.

To calculate the velocities in a narrow temperature range from 0 to 50 keV, a formula can be used that yields less than 1% of the error:

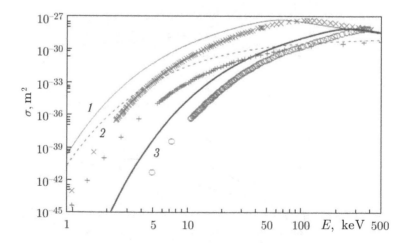

Fig. 6.2. The reaction cross section: D–T – curve *1* [81], × [345, 346]; D–D – curve *2* [81], + [345, 346]; D–³He – curve *3* [81], o [345, 346].

$$\langle\sigma v\rangle = (A0 + A1{\cdot}T + A2{\cdot}T^2 + A3{\cdot}T^3 + A4{\cdot}T^4 + A5{\cdot}T^5){\cdot}10^{-30}, \qquad (6.1)$$

The corresponding coefficients are shown in Table 26. The error in comparison with the last experimental data is shown in the last column. The improved dependence for the D–³He-reaction is presented in the right column.

Below are given formulas for the cross sections and rates of different reactions in different temperature ranges. Table 27 contains coefficients for reaction rates over the entire energy range, Tables 28 and 29 include approximation coefficients for six and three reactions, respectively.

The approximation formula for reaction cross sections:

$$\sigma(T) = A6{\cdot}T^6 A5{\cdot}T^5 + A4{\cdot}T^4 + A3{\cdot}T^3 + A{\cdot}T^2 + A1{\cdot}T^1 + A0.$$

Table 25. Coefficients for the cross sections of the main reactions

Coefficient	$T(d, n)^4$He		^{3}He$(d, p)^4$He		
$A0$	$2.98039 \cdot 10^{-6}$	$1.87311 \cdot 10^{-4}$	$-7.47941 \cdot 10^{-5}$	$-9.992065 \cdot 10^{-2}$	
$A1$	$-3.00864 \cdot 10^{-6}$	$-5.96097 \cdot 10^{-5}$	$1.89346 \cdot 10^{-5}$	$2.93833 \cdot 10^{-2}$	
$A2$	$9.21275 \cdot 10^{-7}$	$3.82890 \cdot 10^{-6}$	$-1.34338 \cdot 10^{-6}$	$-3.26625 \cdot 10^{-5}$	
$A3$	$-1.34859 \cdot 10^{-7}$	$-2.16881 \cdot 10^{-8}$	$2.75544 \cdot 10^{-8}$	$1.76810 \cdot 10^{-7}$	
$A4$	$9.31266 \cdot 10^{-9}$	$4.84185 \cdot 10^{-11}$	$7.44298 \cdot 10^{-12}$	$-2.91064 \cdot 10^{-10}$	
$A5$	$-1.43407 \cdot 10^{-10}$	$-3.79223 \cdot 10^{-14}$	$-5.67748 \cdot 10^{-14}$	$1.05051 \cdot 10^{-13}$	
Range E, keV	16–70	70–100	32–46	46–125	125–400
Error, %	1.9	6.6	9.3	5.2	1.5
Coefficient	$D(d, p)$T		$D(d, n)^3$He		
$A0$	$4.18979 \cdot 10^{-7}$	$-8.884352 \cdot 10^{-4}$	$4.88723 \cdot 10^{-6}$	$6.09512 \cdot 10^{-4}$	
$A1$	$-7.74547 \cdot 10^{-7}$	$-6.59268 \cdot 10^{-6}$	$-4.43152 \cdot 10^{-6}$	$-1.12732 \cdot 10^{-4}$	
$A2$	$4.32851 \cdot 10^{-7}$	$3.06213 \cdot 10^{-6}$	$1.45622 \cdot 10^{-6}$	$5.29133 \cdot 10^{-6}$	
$A3$	$-9.48211 \cdot 10^{-8}$	$-1.73700 \cdot 10^{-8}$	$-2.21348 \cdot 10^{-7}$	$-3.46232 \cdot 10^{-8}$	
$A4$	$8.00031 \cdot 10^{-9}$	$3.82478 \cdot 10^{-11}$	$1.49976 \cdot 10^{-8}$	$1.00089 \cdot 10^{-10}$	
$A5$	$-1.28587 \cdot 10^{-10}$	$-2.95436 \cdot 10^{-14}$	$-2.74461 \cdot 10^{-10}$	$-1.09585 \cdot 10^{-13}$	
Range E, keV	6–27	27–40	40–400	7–22	22–300
Error, max,%	4.6	9.3	4.8	5.4	4.3

Table 26. Approximation coefficients for reaction rates

Coefficient	Reaction		
	D + T → n + ⁴He		D + ³He → p + ⁴He
$A0$	$-1.20006 \cdot 10^6$	$2.16487 \cdot 10^7$	$3.63902 \cdot 10^{-1}$
$A1$	$2.40784 \cdot 10^6$	$-2.30922 \cdot 10^7$	$-1.0457 \cdot 10^{-3}$
$A2$	$-1.67140 \cdot 10^6$	$4.74154 \cdot 10^6$	$8.52512 \cdot 10^{-4}$
$A3$	$4.66463 \cdot 10^5$	$-1.78174 \cdot 10^5$	$6.17099 \cdot 10^{-6}$
$A4$	$-2.41008 \cdot 10^4$	$2.82572 \cdot 10^3$	$-3.1121 \cdot 10^{-8}$
$A5$	$3.10873 \cdot 10^2$	$-1.68168 \cdot 10^1$	$2.04725 \cdot 10^{-11}$
T, keV	4–9	9–50	8–65

Coefficient	Reaction		
	D + D → n + ³He		D + D → p + T
$A0$	$-2.17734 \cdot 10^3$	$1.35366 \cdot 10^5$	$-2.06727 \cdot 10^2$
$A1$	$5.93451 \cdot 10^3$	$-7.87975 \cdot 10^4$	$5.80139 \cdot 10^3$
$A2$	$-5.98913 \cdot 10^3$	$1.54469 \cdot 10^4$	$-6.03022 \cdot 10^3$
$A3$	$2.49654 \cdot 10^3$	$-3.34032 \cdot 10^2$	$2.59034 \cdot 10^3$
$A4$	$-1.91944 \cdot 10^2$	3.91717	$-2.11411 \cdot 10^2$
$A5$	5.62182	$-1.948849 \cdot 10^{-2}$	6.45466
T, keV	2–8	8–50	2–9

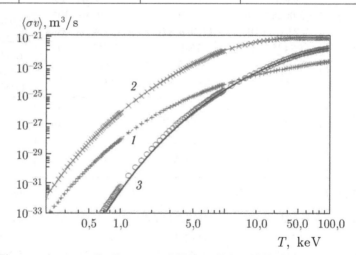

Fig. 6.3. The reaction rate: D–T – curve *1* [81], × [345, 346]; D–D – curve *2* [81], + [345, 346]; D–³He – curve *3* [81], o [345, 346].

The approximation formula for the reaction velocity sections in a narrower range:

$$\langle \sigma v \rangle (T) = A6 \cdot T^6 A5 \cdot T^5 + A4 \cdot T^4 + A3 \cdot T^3 + A2 \cdot T^2 + A1 \cdot T^1 + A0.$$

The safety and attractiveness of thermonuclear fusion systems can be improved with the use of neutrinuclear fuels. Two fuel fusion cycles based on the ^{3}He(d, p)^{4}He and ^{11}B(p, $\alpha\alpha$) α-reactions (the D–^{3} He reaction is not completely neutron-free, hence we will call these cycles slightly radioactive) are interesting both for alternative systems, and from the point of view of non-electrical applications of plasma.

The reaction scheme D–^{3}He: in the collision of deuterium and helium, an α-particle with an energy of 3.67 MeV and a proton with an energy of 14.68 MeV are formed. The maximum reaction cross section is 0.71 b (1 b = 10^{24} cm^2 = 10^{28} m^2) at a colliding particle energy of 470 keV.

The scheme of the reaction p–^{11}B: when a proton collides with the ^{11}B nucleus, an α-particle with an energy of 4.1 MeV and a core of ^{8}Be are formed, which decays in 10^{-16} s into 2 α-particles with an energy of 2.29 MeV each. The maximum cross section of the reaction is observed at an energy of colliding particles of 590 keV (resonance cross section 1.2b), which corresponds to a strict dependence of the Coulomb barrier permeability factor (Gamow's function) on energy. The polarization of nuclei increases this cross section by a factor of 1.6 [314].

The p–^{11}B-reaction cross section has two distinct peaks. Of greatest interest from the point of view of applied research is the resonance at 148 keV, in which the cross section is 0.2 b. The linear scale is chosen to show the dependence of the cross section on the energy in the centre-of-mass system (Figure 6.4).

Approximation of the velocities of neutron-free reactions looks like

$$\langle \sigma v \rangle = (A0 + A1{\cdot}T + A2{\cdot}T^2 + A3{\cdot}T^3 + A4{\cdot}T^4 + A5{\cdot}T^5){\cdot}T{\cdot}10^{-30}. \qquad (6.2)$$

Table 27. Approximation coefficients for the reaction cross sections

Coefficient	Reaction			
	$D + T \rightarrow n + {}^4He$		$D + 3\,He \rightarrow p + {}^4He$	
$A0$	$-9.348095 \cdot 10^{-1}$	$1.403580 \cdot 10^4$	$-1.509100 \cdot 10^{-1}$	$-7.937354 \cdot 10^3$
$A1$	$1.404933 \cdot 10^0$	$-1.944736 \cdot 10^2$	$1.668581 \cdot 10^{-1}$	$1.076491 \cdot 10^2$
$A2$	$-7.604202 \cdot 10^{-1}$	$1.231263 \cdot 10^0$	$-1.442096 \cdot 10^{-2}$	$-5.133487 \cdot 10^{-1}$
$A3$	$9.061945 \cdot 10^{-2}$	$-4.241757 \cdot 10^{-3}$	$3.907267 \cdot 10^{-4}$	$1.223852 \cdot 10^{-3}$
$A4$	$4.101276 \cdot 10^{-4}$	$8.198784 \cdot 10^{-6}$	$-2.420168 \cdot 10^{-6}$	$-1.578178 \cdot 10^{-6}$
$A5$	$-3.976209 \cdot 10^{-5}$	$-8.358801 \cdot 10^{-9}$	$1.018697 \cdot 10^{-8}$	$1.054174 \cdot 10^{-9}$
$A6$	$2.931605 \cdot 10^{-7}$	$3.500489 \cdot 10^{-12}$	$-2.083034 \cdot 10^{-11}$	$-2.864307 \cdot 10^{-13}$
T, keV	0–73	73–550	0–190	190–890

Coefficient	Reaction			
	$D + D \rightarrow n + {}^3He$		$D + D \rightarrow p + T$	
$A0$	$-5.861616 \cdot 10^{-1}$	$8.044778 \cdot 10^1$	$-3.774519 \cdot 10^{-1}$	$6.004241 \cdot 10^1$
$A1$	$4.695915 \cdot 10^{-1}$	$6.401269 \cdot 10^{-2}$	$4.312066 \cdot 10^{-1}$	$5.140286 \cdot 10^{-2}$
$A2$	$-1.009333 \cdot 10^{-3}$	$-5.612099 \cdot 10^{-5}$	$-1.142508 \cdot 10^{-3}$	$-3.225673 \cdot 10^{-5}$
$A3$	$1.291237 \cdot 10^{-6}$	$2.110903 \cdot 10^{-8}$	$1.788772 \cdot 10^{-6}$	$9.832125 \cdot 10{-9}$
$A4$	$-9.733082 \cdot 10^{-10}$	$-4.105646 \cdot 10^{-12}$	$-1.580736 \cdot 10^{-9}$	$-1.769433 \cdot 10^{-12}$
$A5$	$3.927302 \cdot 10^{-13}$	$4.115361 \cdot 10^{-16}$	$7.244732 \cdot 10^{-13}$	$1.808455 \cdot 10^{-16}$
$A6$	$-6.514962 \cdot 10^{-17}$	$-1.768229 \cdot 10^{-20}$	$-1.336237 \cdot 10^{-16}$	$-7.858607 \cdot 10^{-21}$
T, keV	0–1490	1490–4900	0–1490	1490–5000

Coefficient	Reaction		
	$p + {}^6Li \rightarrow {}^4He + {}^3He$		
$A0$	$1.411494 \cdot 10^{-3}$	$5.684833 \cdot 10^0$	$-2.059349 \cdot 10^2$
$A1$	$-1.138507 \cdot 10^{-4}$	$-2.140949 \cdot 10^{-2}$	$5.351635 \cdot 10^{-1}$
$A2$	$1.656044 \cdot 10{-6}$	$3.108116 \cdot 10{-5}$	$-5.763708 \cdot 10^{-4}$
$A3$	$-3.296549 \cdot 10^{-9}$	$-1.998662 \cdot 10^{-8}$	$3.300834 \cdot 10^{-7}$
$A4$	$2.112413 \cdot 10^{-12}$	$3.942848 \cdot 10^{-12}$	$-1.061424 \cdot 10^{-10}$
$A5$	$2.879698 \cdot 10^{-16}$	$1.280738 \cdot 10^{-15}$	$1.818612 \cdot 10^{-14}$
$A6$	$-5.473635 \cdot 10^{-19}$	$-4.898863 \cdot 10^{-19}$	$-1.297895 \cdot 10^{-18}$
T, keV	0–940	940–1760	1760–2500

Tables 30–32 show the coefficients obtained for the non-basic but already very interesting thermonuclear fusion reactions, such as p–${}^{11}B$ and p–${}^6Li$ (Table 30), T–T (Table 31) and ${}^3He$–${}^3He$ (Table 32).

Table 28. Coefficients for rapid cross sections of reactions

Coefficient	Reaction			
	D + T → n + ^{4}He		D + ^{3}He → p + ^{4}He	
$A0$	$-2.267698 \cdot 10^{-18}$	$-5.659726 \cdot 10^{-16}$	0	$-1.604238 \cdot 10^{-16}$
$A1$	$4.256077 \cdot 10^{-18}$	$7.394275 \cdot 10^{-17}$	$7.281172 \cdot 10^{-20}$	$4.475286 \cdot 10^{-18}$
$A2$	$-2.629279 \cdot 10^{-18}$	$-1.449199 \cdot 10^{-18}$	$-1.842297 \cdot 10^{-20}$	$8.045085 \cdot 10^{-21}$
$A3$	$6.905971 \cdot 10^{-19}$	$1.325856 \cdot 10^{-20}$	$1.427949 \cdot 10^{-21}$	$-4.124189 \cdot 10^{-22}$
$A4$	$-4.866666 \cdot 10^{-20}$	$-4.372883 \cdot 10^{-23}$	$-1.209357 \cdot 10^{-23}$	$2.965391 \cdot 10^{-24}$
$A5$	$1.500782 \cdot 10^{-21}$	$-1.360374 \cdot 10^{-25}$	$-4.719027 \cdot 10^{-26}$	$-9.010594 \cdot 10^{-27}$
$A6$	$-1.779171 \cdot 10^{-23}$	$9.834630 \cdot 10^{-28}$	$6.247687 \cdot 10^{-28}$	$1.027195 \cdot 10^{-29}$
T, keV	0–20	20–100	0–80	80–190

Coefficient	Reaction	
	D + D → n + ^{3}He	D + D → p + T
$A0$	$4.471919 \cdot 10^{-20}$	$3.083524 \cdot 10^{-20}$
$A1$	$-5.103492 \cdot 10^{-20}$	$-4.317405 \cdot 10^{-20}$
$A2$	$1.309424 \cdot 10^{-20}$	$1.226981 \cdot 10^{-20}$
$A3$	$-2.615993 \cdot 10^{-22}$	$-2.689626 \cdot 10^{-22}$
$A4$	$3.040699 \cdot 10^{-24}$	$3.335635 \cdot 10^{-24}$
$A5$	$-1.914569 \cdot 10^{-26}$	$-2.193142 \cdot 10^{-26}$
$A6$	$4.993384 \cdot 10^{-29}$	$5.875634 \cdot 10^{-29}$
T, keV	0–100	0–100

Coefficient	Reaction			
	p + ^{6}Li → ^{4}He + ^{3}He		p + ^{11}B → 3 ^{4}He	
$A0$	$1.974099 \cdot 10^{4}$	$1.831152 \cdot 10^{7}$	$-1.304491 \cdot 10^{-25}$	$-4.227238 \cdot 10^{-22}$
$A1$	$-1.188691 \cdot 10^{4}$	$-1.184916 \cdot 10^{5}$	$1.367721 \cdot 10^{-25}$	$6.059965 \cdot 10^{-24}$
$A2$	$4.719796 \cdot 10^{2}$	$1.871117 \cdot 10^{3}$	$-1.002235 \cdot 10^{-26}$	$-1.868953 \cdot 10^{-26}$
$A3$	$4.134672 \cdot 10^{1}$	$-3.507684 \cdot 10^{0}$	$2.184590 \cdot 10^{-28}$	$3.034687 \cdot 10^{-29}$
$A4$	$-4.514710 \cdot 10^{-1}$	$2.997395 \cdot 10^{-3}$	$-3.333108 \cdot 10^{-31}$	$-2.770870 \cdot 10^{-32}$
$A5$	$1.871535 \cdot 10^{-3}$	$-1.230786 \cdot 10^{-6}$	$-5.351908 \cdot 10^{-33}$	$1.344648 \cdot 10^{-35}$
$A6$	$-2.783996 \cdot 10^{-6}$	$1.896373 \cdot 10^{-10}$	$1.769697 \cdot 10^{-35}$	$-2.695725 \cdot 10^{-39}$
T, keV	0–220	220–1000	0–170	170–1000

Table 29. Approximating coefficients for the reaction velocity sections (for the entire range)

Coefficient	Reaction		
	D + T → n + ⁴He	D + ³He → p + ⁴He	p + ⁶Li → ⁴He + ³He
$A0$	$3.462196 \cdot 10^{-18}$	$3.214362 \cdot 10^{-18}$	$-1.340452 \cdot 10^{5}$
$A1$	$-1.188011 \cdot 10^{-17}$	$-9.245847 \cdot 10^{-19}$	$6.135338 \cdot 10^{3}$
$A2$	$3.436025 \cdot 10^{-18}$	$4.385183 \cdot 10^{-20}$	$1.858520 \cdot 10^{3}$
$A3$	$-1.228141 \cdot 10^{-19}$	$1.558703 \cdot 10^{-22}$	$-4.983392 \cdot 10^{0}$
$A4$	$1.937993 \cdot 10^{-21}$	$-6.659560 \cdot 10^{-24}$	$6.858470 \cdot 10^{-3}$
$A5$	$-1.460953 \cdot 10^{-23}$	$4.139431 \cdot 10^{-26}$	$-5.031526 \cdot 10^{-6}$
$A6$	$4.279513 \cdot 10^{-26}$	$-8.044398 \cdot 10^{-29}$	$1.512241 \cdot 10^{-9}$
T, keV	0–100	0–190	0–1000

Table 30. Coefficients for reaction rates of p–¹¹B and p–⁶Li

Coefficient	Reaction		
	p + ¹¹B → 3 ⁴He	p + ⁶Li → ⁴He + ³He	
$A0$	$6.83699 \cdot 10^{5}$	$7.38975 \cdot 10^{3}$	$-2.26957 \cdot 10^{1}$
$A1$	$8.0053 \cdot 10^{3}$	$-1.65358 \cdot 10^{3}$	$1.92015 \cdot 10^{3}$
$A2$	$-3.63029 \cdot 10^{1}$	$1.17234 \cdot 10^{2}$	-5.22401
$A3$	$6.11443 \cdot 10^{-2}$	-1.64418	$7.29619 \cdot 10^{-3}$
$A4$	$-4.69081 \cdot 10^{-5}$	$1.05352 \cdot 10^{-2}$	$-5.40516 \cdot 10^{-6}$
$A5$	$1.37171 \cdot 10^{-8}$	$-2.67543 \cdot 10^{-5}$	$1.63323 \cdot 10^{-9}$
T, keV	200–1000	10–130	130–1000

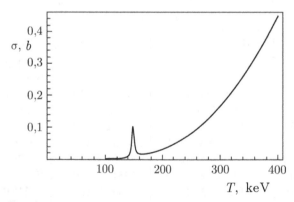

Fig. 6.4. The effective cross section for the ¹¹B(p, αα)α reaction.

A comparison of the latest data [347] and earlier experiments [345, 346] shows that in the range from 100 to 300 keV, the most interesting for thermonuclear fusion, there is a difference of ~30% in the rate of the p–¹¹B fusion reaction. Figure 6.5 shows the low-

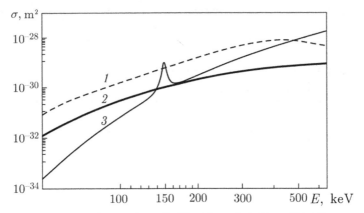

Fig. 6.5. The sections D–³He (*1*), p–⁶Li (*2*), p–¹¹B (*3*).

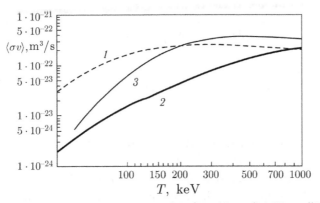

Fig. 6.6. The reaction rates of D–³He (*1*), p–⁶Li (*2*), p–¹¹B (*3*).

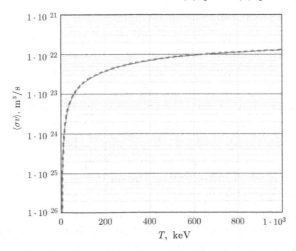

Fig. 6.7. Dependence of the reaction rate on temperature for T + T → 2n + ⁴He (1–1000 keV). Approximate dependence (dashed line), data FEDNL2.0 (solid line).

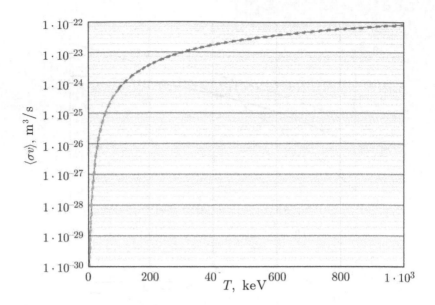

Fig. 6.8. Dependence of the reaction rate of ^{3}He + ^{3}He on the temperature in the range 20–1000 keV (dashed line). The AEP data is a solid line.

radioactive reactions and their effective cross sections. The velocities of these reactions are shown in Fig. 6.6.

Combine all the reaction rates on one graph (Fig. 6.9). In view of the small differences in the approximate values of the reaction rate from the actual values (taken from FENDL 2.0, AEP, Newins, ECPL-82), these formulas can be used.

Now let us turn to the topic of polarized atoms. Although this question applies more to technology, it should be noted that the polarization of thermonuclear fuel increases σ by 1.5 times for D–T and D–^{3}He and by 1.6 times for the p^{-11} B reaction.

The polarization of nuclear spins using the technique of optical pumping to atoms has a long history [348]. Polarization is well studied theoretically and has been applied to a large number of atoms. A classic review on this topic [349] even after 50 years is a fundamental work on this topic. The latest relevant monograph was published by Sater [350].

Polarized ions and atomic beams have been extensively used in nuclear physics, and it has been established that laser optical pumping is the best method under the conditions where it is applicable. Simplicity, speed and reachability of high polarization make it the most attractive of all known methods. Milliampere currents (about

Table 31. Coefficients for the reaction rate T–T

Coefficient	Reaction	
	$T + T \rightarrow 2n + {}^4He$	
$A0$	$1.934579 \cdot 10^4$	$-2.025050 \cdot 10^6$
$A1$	$-3.875392 \cdot 10^4$	$2.167020 \cdot 10^5$
$A2$	$9.366280 \cdot 10^3$	$-6.931906 \cdot 10^1$
$A3$	$-1.666830 \cdot 10^2$	$-4.236486 \cdot 10^{-2}$
$A4$	1.528408	$-1.649827 \cdot 10^{-5}$
$A5$	$-6.867526 \cdot 10^{-3}$	$7.932323 \cdot 10^{-8}$
$A6$	$1.162797 \cdot 10^{-5}$	$-3.748199 \cdot 10^{-11}$
T, keV	$1-125$	$125-1000$

Table 32. The coefficients for the reaction rate of ${}^3He-{}^3He$

Coefficient	Reaction		
	${}^3He + {}^3He \rightarrow 2p + {}^4He$		
$A0$	$-7.11688 \cdot 10^{-1}$	$-9.41344 \cdot 10^3$	$4.08451 \cdot 10^5$
$A1$	1.10871	$1.35306 \cdot 10^3$	$-1.45149 \cdot 10^4$
$A2$	$-6.02460 \cdot 10^{-1}$	$-6.93360 \cdot 10^1$	$1.68045 \cdot 10^2$
$A3$	$1.50260 \cdot 10^{-1}$	1.24727	$-2.72911 \cdot 10^{-2}$
$A4$	$-1.83835 \cdot 10^{-2}$	$5.55646 \cdot 10^{-3}$	$-1.41085 \cdot 10^{-4}$
$A5$	$9.88063 \cdot 10^{-4}$	$-9.55992 \cdot 10^{-5}$	$1.33599 \cdot 10^{-7}$
$A6$	$-1.06198 \cdot 10^{-5}$	$3.13005 \cdot 10^{-7}$	$-3.75770 \cdot 10^{-11}$
T, keV	$6-20$	$20-100$	$100-1000$

10^{16}/s) can generate H, D, ^{3}He, ^{6}Li and ^{7}Li with a high polarization coefficient and produce their ions continuously. Figure 6.10 shows the rates of ordinary reactions in comparison with resonant ones.

It was shown in [351, 352] that an increase in current by 2 orders of magnitude (at least up to $4 \cdot 10^{17}$/s) for the polarization of hydrogen is achieved by using the so-called exchange optical pumping. The required requirement is 10^{20}/s of the spin of polarized hydrogen atoms. The power of lasers should reach several hundred kW, which is rather modest by modern concepts.

Polarization effects can have a significant effect on the cross section of the reactions, while the experimental data of the reactions are mainly measured with unpolarized colliding nuclei [353, 354].

Theoretical calculations of σ are extremely complicated, and the accuracy of such calculations is not guaranteed [355]. Only an experimental measurement can give a definite answer. It is very important to measure the reaction cross sections for the case of two polarized initial nuclei.

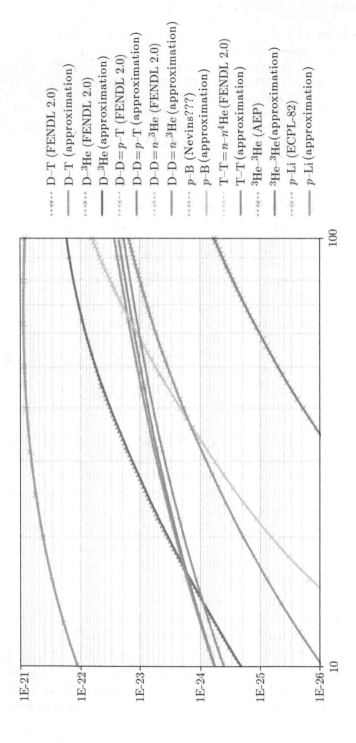

Fig. 6.9. Comparison of approximation dependences (solid lines) with experimental data (dashed lines) for the rates of different fusion reactions $\langle \sigma v \rangle$ (m³/s) versus temperature T (keV).

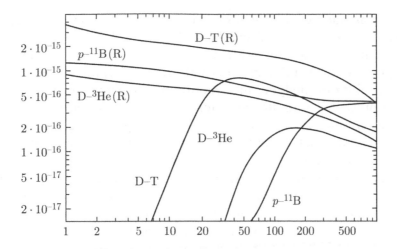

Fig. 6.10. The reaction rates (σv, 10^{-6} m³/s) thermallized and resonant (R on temperature (T, keV)

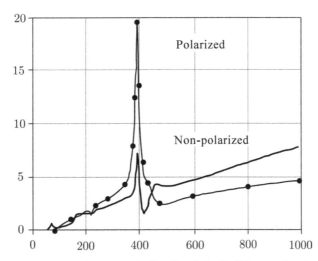

Fig. 6.11. The cross sections (σ, 10^{-27} m²) of the D–³He-reaction in the resonance region as a function of energy (E, keV).

To date, experimental cross-section measurements have been performed for the cases when only one particle of the two is polarized. Figure 6.11 shows the dependence of the D–³He-reaction cross section on energy for polarized and unpolarized atoms. In this reaction, the formation of an intermediate ⁵Li nucleus provides an increase in the resonance cross section [356]. The factor of the relative increase in the effective cross section in the resonance region was adopted in the ARTEMIS project 1.5 times the cross section for the reaction of unpolarized atoms [357]. Since there are no experimental data for cross-section measurements, it would be very good to measure them. This will give a definite answer about the possibility of applying the reaction at low temperatures.

The presented formulas are more simplified, which reduces the calculation time. The D–^{3}He reaction is most promising for use in a nuclear fusion reactor due to the absence of neutron radiation, and in the case of the p–^{11}B reaction, a discrepancy of σ and especially $\langle \sigma v \rangle$ with early formulas in the entire energy range is obvious. The above analytical dependences can be used to calculate thermonuclear plasma over a wide energy range, and the results obtained are very useful for further studies of the problems of thermonuclear fusion.

References

1. *Velikhov E.P., Smirnov V.P.,* The state of research and the prospect of thermonuclear energy, VANT. Ser. Termoyadernyi sintez. 2006. Issue 4. P. 3–14.
2. ITER Physics Basis, Nucl. Fusion. 1999. V. 39. P. 2137–2638.
3. *Kolbasov B.N., Borisov A.A., Vasiliev N.N., et al.* The concept of the demonuclear thermonuclear power reactor DEMO-S. Problems of Atomic Science and Technology. Series Thermonuclear Fusion, 2007. Issue 4. P. 3–13.
4. *Stacey W.M.* Tokamak D–T fusion neutron source requirements for closing the nuclear fuel cycle, Nucl. Fusion. 2007. V. 47. P. 217–221.
5. *Kuteev B.V., Goncharov P. P., Sergeev V.Yu., Khripunov V.I.,* Powerful neutron sources based on the nuclear fusion reaction, Plasma Physics. 2010. P. 36. P. 307–346.
6. *Golovin I.N., Kostenko V.V., Khvesyuk V.I., Shabrov N.V.,* Estimation of plasma parameters of a thermonuclear reactor on D^3He-fuel, Pis'ma v ZhTF. 1988. Vol. 14, No. 20. P. 1860–1863.
7. *Kurtmullaev R.Kh., Malyutin A.I., Semenov V.N.,* Compact torus, Itogi Nauki i Tekhniki. Fizika plazmy. Moscow: VINITI, 1985. V. 7. No. 7. P. 80–135.
8. *Slough J.T.* Suitability of small scale linear systems for a fission-fusion reactor, breeder, and waste transmutation. J. Fusion Energy. 2008. V. 27. P. 115–118.
9. *Golovin I. N.,* Low-radioactive controlled thermonuclear fusion (reactors with $D-^3He$). Moscow, 1989. 49 p. (Preprint IAE No. 4885/8).
10. *Santarius J.F., Kulcinski G.L., El-Guebaly L.A., Khater H.Y.,* Could advanced fusion fuels be used with today's technology?, J. Fusion Energy. 1998. V. 17. P. 33–40.
11. *Khvesyuk V.I., Chirkov A.Yu.,* Low-radioactivity $D-^3He$ fusion fuel cycles with ^3He production, Plasma Phys. Control. Fusion. 2002. V. 44. P. 253–260.
12. *Khvesyuk V.I, Chirkov A.Yu.* Low-radioactive $D-^3He$ thermonuclear fuel cycle with ^3He self-supply. Tech, Phys. Letters. 2001. V. 27, No. 8. P. 686–688.
13. *Khvesyuk V.I., Shabrov N.V.,* To the problem of transverse pumping of ions from a mirror cell by an external rotating magnetic field, Pis'ma v ZhTF. 1993. V 19. P. 42–45.
14. *Putvinsky S.V.,* Alfa particles in a tokamak, in: Problems in the Theory of Plasma. Issue 18, Ed. B.B. Kadomtsev. Moscow: Energoatomizdat, 1990. P. 209–315.
15. *Wurden G.A., Hsu S.C., Intrator T.P., et al.* Magneto-Inertial Fusion, J. Fus. Energy. 2016. V. 35. P. 69–77.
16. URL http://www.helionenergy.com/Helion_Presentation-Web2.pdf.
17. http://www.generalfusion.com/
18. *Bussard R.,* The advent of clean nuclear fusion: superperformance space power and propulsion, 57th International Astronautical Congress. International Astronautical Federation. 2006. V. 57. URL: http://www.emc2fusion.org/
19. *Galimov E.M, Anufriev G.S.,* He-3 in the lunar soil in the depth of the column selected by the automatic station Luna-24, Doklady Akademii Nauk.2007. V. 412, No. 3. P. 388–390.

20.   *Li D.H., et al.* Lunar ³He estimates and related parameters analyses, Chinese Acad-
       emy of Sciences, Science China – Earth Sciences. 2010. V. 53. P. 1103–1114.
21.   *Santarius J.F., Kulcinski G.L., El-Guebaly L.A.,* A passively proliferation-resistant
       fusion power plant. Fusion Science and Technology. 2003. V. 44. P. 289–293.
22.   *Kuteev B.V., Khripunov V.I..* Modern view of the hybrid thermonuclear reactor,
       Problems of Atomic Science and Technology. Series Thermonuclear Fusion. 2009.
       No. 1. P. 3–29.
23.   *Khvesyuk V.I,, Chirkov A.Yu.,* Analysis of fuel cycles for alternative thermonuclear
       reactors, Problems of Atomic Science and Technology. Series Thermonuclear Fu-
       sion. 2000. Issue 3. P. 28–35.
24.   *Ryzhkov S.V.* Modeling of plasma physics in the fusion reactor based on a field-
       reversed configuration, Fusion Science and Technology. 2009. V. 55, No. 2T. P.
       157–161.
25.   *Chirkov A.Yu., Khvesyuk V.I.,* To the calculation of the distribution functions of high-
       energy ions in terms of velocities, Problems of Atomic Science and Technology.
       Series Thermonuclear Fusion. 2003. Issue.1. P. 55–65.
26.   *Khvesyuk V.I., Chirkov A.Yu.,* Analysis of the mechanisms for the scattering of plas-
       ma particles by non-steady-state fluctuations, Tech,. Phys. 2004. V. 79, No. 4. P.
       396–404.
27.   *Chirkov A.Yu., Khvesyuk V.I.,* Electromagnetic drift instabilities in high-β plasma
       under conditions of a field reversed configuration, Phys. Plasmas. 2010. V. 17, No.
       1. 012105.
28.   *Chirkov A.Yu.,* On the possible concept of a tokamak reactor with an alternative
       thermonuclear cycle D–³He, Problems of Atomic Science and Technology. Series
       Thermonuclear Fusion. 2001. Issue 2. P. 37–43.
29.   *Chirkov A.*Yu., On the possibility of using a D–³He-cycle with ³He production in a
       thermonuclear reactor based on a spherical tokamak, Zh. Teor. Fiz. 2006. V. 76, No.
       9. P. 51–54.
30.   *Rudakov B.A.* Reactor-stellarator based on D–³He-synthesis. Reports of the Confer-
       ence on Low-Radioactive Thermonuclear Fusion based on D–³He. Moscow: IAE,
       1991. WP. 93.
31.   *Chirkov A.Yu.,* Estimation of plasma parameters in a D–³He-reactor based on an
       inverse magnetic configuration, Problems of Atomic Science and Technology. Series
       Thermonuclear Fusion. 2006. Issue. 4. P. 57–67.
32.   *Ryzhkov S.V.,* Comparison of a deuterium–helium-3 FRC and mirror trap for plasma
       confinement, Fusion Science and Technology. 2007. V. 51, No. 1T. P. 190–192.
33.   *Chirkov A.Yu.,* Low Radioactivity Fusion Reactor Based on the Spherical Tokamak
       with a Strong Magnetic Field, J. Fusion Energy. 2013. V. 32. P. 208–214.
34.   *Volkov E.D., Rudakov V.A., Suprunenko V.A.,* Optimization of the ignition scenario
       in the torsatron reactor, Engineering problems of thermonuclear reactors: Proc. Sec-
       ond All-Union. Conf. Leningrad: NIIEFA, 1981. P. 173.
35.   *Ryzhkov S.V.,* Compact toroid and advanced fuel – together to the Moon?!, Fusion
       Science and Technology. 2005. V. 47, No. 1T. P. 342–344.
36.   *Chirkov A.Yu.,* Alternative systems of thermonuclear synthesis. Moscow: Kniga i
       biznes, 2012.
37.   *Ryzhkov S.V.,* Low radioactive and hybrid fusion – a path to clean energy, Sustain-
       able Cities and Society. 2014. V. 15. P. 313–315.
38.   *Ryzhkov S.V.,* Reversed magnetic configuration and applications of high-temperature
       plasma FRC, Applied Physics. 2010. No. 1. P. 47–54.
39.   *Ryzhkov SV* Modeling of thermophysical processes in a magnetic thermonuclear

engine, Thermal processes in engineering. 2009. № 9. P. 397–400.

40. *Kulcinski G.L., et al.,* Summary of APOLLO, a D–³He tokamak reactor design, Fusion Technology. 1992. V. 21. P. 2292–2315.

41. *Bathke et al.* Systems analysis in support of the selection of the ARIES-RS design point, Fusion Eng. and Design. 1997. V. 38. P. 59–74.

42. *Momota H., et al.* Conceptual design of D–³He FRC reactor ARTEMIS, Fusion Technology. 1992. V. 21. P. 2307–2323.

43. *Ryzhkov S.V., Khvesyuk V.I., Ivanov A.A.* Progress in an alternate confinement system called a FRC, Fusion Science and Technology. 2003. V. 43, No. 1T. P. 304–308.

44. *Stambaugh R.D., et al.* The spherical tokamak path to fusion power, Fusion Technology. 1998. V. 33, No. 1. P. 1–21.

45. *Moir R.W., et al.,* Thick liquid-walled, field-reversed configuration magnetic fusion power plant, Fusion Technology. 2001. V. 39, No. 2. P. 758–767.

46. *Vershkov V.A.,* Anomalous transport and small-scale turbulence in a tokamak: Dissertation Dr. Phys.-Math. Sciences. Moscow, 2009. 292 p.

47. *Moir R.W.* Liquid first walls for magnetic fusion energy configurations, Nuclear Fusion. 1997. V. 37. P. 557–566.

48. *Khvesyuk V.I., Ryzhkov S.V., Santarius J.F., et al.,* D–³He reversed configuration fusion power plant, Fusion Technol. V. 39. No. 1T. 2001. P. 410–413.

49. URL: http://www.iter.org/Parameters.htm (circulation date 09.10.2011)

50. *Pratt J., Horton W.,* Global energy confinement scaling predictions for the kinetically stabilized tandem mirror, Phys. Plasmas. 2006. V. 13. 042513 (9 p.).

51. *Kulygin V.M., et al.,* On the way to project Epsilon, Nucl. Fusion. 2007. V. 47. P. 738–745.

52. *Kesner J., et al.,* Helium catalyzed D–D fusion in a levitated dipole, Nuclear Fusion. 2004. V. 44. P. 193–203.

53. *Rostoker N., Binderbauer M.W., Monkhorst H.J.,* Colliding Beam Fusion Reactor, Science. 1997. V. 278. P. 1419–1422.

54. *Krivosheev M.V., Izotova A.V., Seko E.V., et al.,* Optimization Estimates for the Parameters of a Thermonuclear Reactor Based on a Long Antiprobe Trap, Vopr. atomnoi nauki i tekhniki. Ser. Termoyadernyi sintez. 1989. Issue 2. P. 29–34.

55. *Voronchev V.T., Kukulin V.I.,* Implementation of the thermonuclear process in D³He-⁹Be plasma on the basis of a Z pinch with an ultrafast laser ignition, Phys. Atom. Nucl. 2010. V. 73. P. 38-58.

56. *Basko M.M., et al.,* Nuclear fusion with inertial confinement. Current state and prospects for energy. Moscow: FIZMATLIT, 2005. 264 p.

57. *Nedoseev, S.L.,* Energy of inertial thermonuclear fusion. The concept of a thermonuclear reactor with a driver based on z-pinch. Moscow: MFTI, 2004. 25 p.

58. *Siemon R.E., et al.* MTF: prospects for low-cost fusion energy, J. Plasma Fusion Res. SERIES. 2002. V. 5. P. 63–68.

59. *Ryzhkov S.V.,* Modeling and engineering applications for weakly turbulent plasma, 35th EPS Conf. on Plasma Physics and Contr. Fusion. Hersonissos, 2008. V. 32D. P 1.114.

60. *Chirkov A.Yu.,* Optimal Parameters of Fusion Neutron Sources with a Powerful Injection Heating, Journal of Fusion Energy. 2015. V. 34. P. 528–531.

61. *Binderbauer M.W., et al.,* Dynamic formation of a hot field reversed configuration with improved confinement by supersonic merging of two colliding high-beta compact toroids, Physical Review Letters. 2010. V. 105. 045003 (4 p.).

62. *Chirkov A.Yu., Khvesyuk V.I.,* Characteristics of collisionless gradient drift instabilities in a plasma with a highly inhomogeneous magnetic field and high β, Fizika

plazmy. 2011. V. 37, No. 5. P. 473–483.

63.   *Khvesyuk V.I.*, General analysis of plasma turbulence and estimation of the excitation of fluctuations by a drift wave, in: Problems of Atomic Science and Technology. Series Thermonuclear Fusion. 2011. Issue 4. P. 72–79.

64.   *Ryzhkov S.V.*, Helium-3 – based fusion plasma, Problems of Atomic Science and Technology. Series: Plasma Physics. 2008. No. 6 (14). P. 61–63.

65.   *Wittenberg L.J., Santarius J.F., Kulcinski G.L.*, Lunar source of ^{3}He for commercial fusion power, Fusion Technology. 1986. V. 10. P. 167–178.

66.   *Kernbichler W.*, Operational parameters for D–^{3}He in fied-reversed configurations, Fusion Technology. 1992. V. 21. P. 2297–2306.

67.   Progress in ITER Physics Basis. Chapter 1, Nucl. Fusion. 2007. V. 47. P. S1–S17.

68.   *Peng Y.-K.M., Strickler D.J.*, Features of spherical torus plasmas, Nucl. Fusion. 1986. V. 26. P. 769–777.

69.   *Sykes A.*, Overview of recent spherical tokamak results, Plasma Phys. Control. Fusion. 2001. V. 43. P. A124–A139.

70.   *Gusev V.K., Golant V.E., Gusakov E.Z., etc.* Spherical Tokamak Globus-M, Zh. Teor. Fiz. 1999. Vol. 69, No. 9. P. 58–62.

71.   *Raman R., Ahn J.-W., Allain J.P. et al.* Overview of physics results from NSTX, Nucl. Fusion. 2011. V. 51. 094011 (18 p.).

72.   *Lloyd B., Akers R. J., Alladio F. et al.* Overview of physics results from MAST, Nucl. Fusion. 2011. V. 51. 094013 (10 p.).

73.   *Galambos J.D., Peng Y.-K.M.*, Ignition and burn criteria for D–^{3}He tokamak and spherical torus reactors, Fusion Technol. 1991. V. 19. P. 31–42.

74.   *Bingren S.*, Core plasma characteristics of a spherical tokamak D–^{3}He fusion reactor, Plasma Sci. Technol. 2005. V. 7. P. 2767–2772.

75.   *Umeda K., Yamazaki K., Oishi T., Arimoto H., Shoji T.* Self-Consistent Pressure and Current Profiles in High-Beta D–^{3}He Tokamak Reactors, J. Plasma Fusion Res. 2010. V. 5. S2030 (4 p.).

76.   *Miyamoto K.*, Fundamentals of plasma physics and controlled thermonuclear fusion. Moscow: FIZMATLIT, 2007. §16.11.

77.   *Shimada M., Chudnovskii A., Costley A., et al.*, Physics Design of ITERFEAT, J. Plasma Fusion Res. SERIES. 2000. V. 3. P. 77–83.

78.   *Subbotin M.L., Kurbatov D.K., Filimonova E.A.*, Review of the state of research of demonstration thermonuclear reactors in the world, VANT. Termoyadernyi sintez. 2010. Issue 3. P. 55–74.

79.   *Stacey W.M.*, Tokamak demonstration reactors, Nucl. Fusion. 1995. V. 35. P. 1369–1384.

80.   *Mirnov S.V., Azizov E.A., Alekseev A.G., et al.*, Li experiments on T-11M and T-10 in support of a steady-state tokamak concept with Li closed loop circulation, Nucl. Fusion. 2011. V. 51. 073044 (9 p.).

81.   *Bosh H.-S., Hale, G. M.* Improved formulas for fusion cross-sections and thermal reactivities, Nucl. Fusion. 1992. V. 32. P. 611–631.

82.   *Chirkov A.Yu.*, Plasma bremsstrahlung emission at electron energy from low to extreme relativistic values, ArXiv e-prints. 2010. arXiv: 1005.3411v1.

83.   *Kukushkin A.B., Minashin P.V., Neverov V.S.*, Electron cyclotron power losses in fusion reactor-grade tokamaks: scaling laws for the profile and power loss, 22nd IAEA Fusion Energy Conf., Geneva, 2008. TH / P3-10.

84.   *Miller R.L.*, System Perspectives of ARIES-ST, Japan-US workshop on fusion power plants and related advanced technology, Kyoto, 1999. P. 185–197.

85. *Valovic M., Akers R., Cunningham G., et al.*, Scaling of H-mode energy confinement with I_p and B_T in the MAST spherical tokamak, Nucl. Fusion. 2009. V. 49. 075016 (8 p).

86. *Dnestrovsky A.Yu., Golikov A.A., Kuteev B.V., Khairutdinov R.R., Gryaznevich M.P.,* Investigation of the steady-state operation of a neutron source based on tokamak, Problems of Atomic Science and Technology. Series Thermonuclear Fusion. 2010. Issue. 4. P. 26–35.

87. *Galvão R. M. O., Ludwig G.O., Del Bosco E., et al.*, Physics and engineering basis of the multi-functional compact tokamak reactor concept, 22nd IAEA Fusion Energy Conf., Geneva, 2008. IAEA-CN-116/FT/P3-20.

88. *Mirnov S.V.*, Do spherical tokamaks have a thermonuclear future?, Fizika plazmy. 2012. P. 38, No. 12. P. 1011–1021.

89. *Azizov E.A., Mineev A.B.*, On the limits of compactness of neutron sources based on tokamak, VANT. Termoyadernyi sintez. 2012. Issue. 2. P. 5–20.

90. *Chirkov A. Yu.*, Energy efficiency of alternative thermonuclear systems with magnetic confinement of plasma, Yadernaya fizika i inzhiniring. 2013. Vol. 4, No. 11–12. P. 1050–1059.

91. *Clemmow F., Dougherty J.*, Electrodynamics of Particles and Plasmas. Trans. from English. Moscow: Mir, 1996.

92. *Chirkov A.Yu., Ryzhkov S.V., Bagryansky P.A., Anikeev A.V.* Plasma kinetics models for fusion systems based on the axially symmetric mirror devices, Fusion Science and Technology. 2011. V. 59, No. 1T. P. 39–42.

93. *Chirkov A.Yu., Ryzhkov S.V., Bagryansky P.A., Anikeev A.V.*, Fusion modes of an axially symmetrical mirror trap with the high power injection of fast particles, Plasma Physics Reports. 2012. V. 38, No. 13. P. 1025–1031.

94. *Anikeev A.V., Bagryansky P.A., Zaitsev K.V., et al.* Energy spectrum of longitudinal losses of ions in GDL during development of Alfvén ion-cyclotron instability, Fizika plazmy. 2015. Vol. 41, No. 10. S. 839–849.

95. *Pastukhov V.P.*, Collisional losses of electrons from an adiabatic trap in a plasma with a positive potential, Nuclear Fusion. 1974. V. 14. P. 3–6.

96. *Cohen R.H., Rensink M.E., Cutler T.A., Mirin A.A.*, Collisional losses of electrostatically confined species in a magnetic mirror, Nuclear Fusion. 1978. V. 18. P. 1229–1243.

97. *Rognlien T.D., Cutler, T.A.*, Transition from Pastukhov to collisional configment in a magnetic and electrostatic well, Nuclear Fusion. 1980. V. 20. P. 1003–1011.

98. *Mirnov V.V., Ryutov D.D.*, Gas-dynamic linear trap for confinement of plasma, Pis'ma v Zh. Teor. Fiz. 1979. Vol. 5, Vol. 11. P. 678–682.

99. *Mirnov V.V., Ryutov D.D.*, Gas-dynamic trap, Voprosy atomnoi nauki i tekhniki. Ser. Termoyadernyi sintez. 1980. Issue. 1. P. 57–66.

100. *Bagryansky P. A., Ivanov A. A., Kruglyakov E.P., et al.*, Gas dynamic trap as high power 14 MeV neutron source, Fusion Engineering and Design. 2004. V. 70. P. 13–33.

101. *Anikeev A.V., Bagryansky P.A., Ivanov A.A., et al.*, Confinement of strongly anisotropic hot-ion plasma in a compact mirror, J. Fusion Energy. 2007. V. 26. P. 103–107.

102. *Dimov G.I.*, The ambipolar trap, Usp. Fiz. Nauk. 2005. V. 175. P. 1185–1206.

103. *Ioffe M.S., Kadomtsev B.B.*, Plasma confinement in adiabatic traps, Usp. Fiz. Nauk V. 100, No. 4. P. 601–639.

104. *Beklemishev A.D., Chaschin M.S.*, Influence of rotation on the stability of plasma in GDL, Fizika plazmy. 2008. V. 34, No. 5. P. 463–472.

105. *Sorokina E.A.*, Global modes of flute instability of a rotating cylindrical plasma, Fizika plazmy. 2009. V. 35, No. 5. P. 472–481.

106. *Mase A., Jeong J.H., Itakura A., et al.* Ambipolar potential effect on a drift-wave

mode in tandem-mirror plasma, Phys. Rev. Lett. 1990. V. 64, No. 19. P. 2281–2284.

107. *Mase A., Itakura A., Inutake M., et al.,* Control of the radial electric field and of turbulent fluctuations in a tandem mirror plasma, Nuclear Fusion. 1991. V. 31, No. 9. P. 1725–1733.

108. *Bogdanov G.F., Golovin I.N., Kucheriaev Yu.A., Pano, D.A.,* Properties of Plasma Formed in Ogre by Injection of a Beam of Molecular Hydrogen Ions, in: Yadernyi sintez. 1962. Appendix. V. 1. P. 215–225.

109. *Damm S.S., Foote J.H., Futch A.N., et al.,* Cooperative effects in a tenuous energetic plasma contained by a magnetic mirror field, Phys. Fluids. 1965. V. 8. P. 1472–1488.

110. *Anikeev A.V., Bagryansky P.A., Ivanov A., Karpushov A.N., et al.,* Ion-hot plasma with high energy content in the gas-dynamic trap. Fizika plazmy. 1999. Vol. 25. P. 499–509.

111. *Khvesyuk V.I., Chirkov A.Yu.,* Energy production in ambipolar reactors with D–T, D–^{3}He and D–D fuel cycles. Tech. Phys. Letters. 2000. V. 26, No. 11. P. 964–966.

112. *Chirkov A.Yu.,* Numerical solution of the Fokker-Planck equation for modeling the modified gas-dynamic plasma regimes in a magnetic trap with heating by intense atomic beams, Fiziko-khimicheskaya kinetika v gazovoi dinamike. 2011. V. 11. www. chemphys.edu.ru/pdf/2011-02-01-029.pdf.

113. *Karney, C.F.F.,* Fokker-Planck and quasilinear codes, Computer Phys. Reports. 1986. V. 4. P. 183–244.

114. *Rosenbluth M.N., MacDonald W.M., Judd D.L.,* Fokker-Planck equation for an in-verse-square force, Phys. Rev. 1957. V. 107. P. 1–6.

115. *Trubnikov B.A.,* Reduction of the kinetic equation in the case of Coulomb collisions to a differential form, Zh. Eksp. Teor. Fiz. 1958. V. 34. S. 1341–1343.

116. *Devaney J. J., Stein M.L.,* Plasma energy deposition from nuclear elastic scattering, Nucl. Sci. Eng. 1971. V. 46. P. 323–333.

117. *Sivukhin D. V.* Coulomb collisions in a fully ionized plasma. Voprosy teorii plazmy. Issue 4, Ed. M. A. Leontovich. Moscow: Atomizdat, 1964. P. 81–187.

118. *Maximov V.V., Anikeev A.V., Bagryansky P. A., et al.,* Spatial profiles of fusion prod-uct flux in the gas dynamic trap with deuterium neutral beam injection, Nucl. Fu-sion. 2004. V. 44. P. 542–547.

119. *Chirkov A.Yu.,* Evaluation of the operational parameters for NBI-driven fusion in low-gain tokamak with two-component plasma, Nucl. Fusion. 2015. V. 55. 113027. (8 p.).

120. *Almagambetov A.N., Chirkov A.Yu.,* Power and Sizes of Tokamak Fusion Neutron Sources with NBI-Enhanced Reaction Rate, Journal of Fusion Energy. 2016. V. 35. P. 845–848.

121. *Tuszewski M., Smirnov A., Deng B.H., et al.* Combined FRC and mirror plasma stud-ies in the C-2 device, Fusion Sci. Technol. 2011. V. 59, No. 1T. P. 23–26.

122. *Kolb A.C., Dobbie C.B., Griem H.R.,* Field mixing and associated neutron produc-tion in a plasma, Physical Review Letters. 1959. V. 3, No. 1. P. 5–7.

123. *Dudnikova G.I., Kurtmullaev R.Kh., Malyutin A.I., Semenov V.N.,* Effect of viscosity on the relaxation of a compact torus, Fizika plazmy. 1989. V. 15. P. 987–991.

124. *Armstrong W.T., et al.* Field-reversed experiments (FRX) on compact toroids, Phys. Fluids. 1981. V. 24, No. 11. P. 2068–2089.

125. *Tuszewski M.* Field reversed configurations, Nuclear Fusion. 1988. V. 28, No. 11. P. 2033–2092.

126. *Rosenbluth M.N., Bussac M.N.,* MHD stability of spheromak, Nuclear Fusion. 1979. V. 19. P. 489–498.

127. *Jarboe T.R.,* Review of spheromak research, Plasma Phys. Control. Fusion. 1994.

V. 36. P. 945–990.
128. *Shafranov V.D.*, On the equilibrium magnetohydrodynamic configurations, Zh. Eksper. Teor. Fiz. 1957. P. 33. P. 710–722.
129. *Dolan T.J.*, Fusion Research: principles, experiments and technology. 2000. 937 p.
130. *Steinhauer L.C. et al.*, FRC 2001: A white paper on FRC development in the next five years, Fusion Technology. 1996. V. 30. P. 116–126.
131. *Ryzhkov S.V.*, Features of the formation, configuration and stability of the field reversed configuration, Problems of Atomic Science and Technology. Series: Plasma Physics. 2002. No. 4 (7). P. 73–75.
132. *Hill M.J.*, On a spherical vortex, Philos. Trans. R. Soc. Ser. A. 1894. Pt. 1, C/XXXV. P. 213–245.
133. *Barnes D.C.*, Profile consistency of an elongated field-reversed configuration. I. Asymptotic theory, Phys. Plasmas. 2001. V. 8, No. 11. P. 4856–4863.
134. *Steinhauer L.C.*, Review of field-reversed configurations, Phys. Plasmas. 2011. V. 18. 070501 (38 p.).
135. *Hoffman A.L., Slough J.T., Steinhauer L.C., et al.*, Field reversed configuration transport. Theory and measurement of flux, energy, and particle lifetimes, Plasma Physics and Controlled Nuclear Fusion Research: Proc. 11th Int. Conf., V. 2, IAEA, Vienna, 1987. P. 541–549.
136. *Rej D.J., Barnes G.A., Baron M.H., et al.* Electron energy confinement in field reversed configuration plasmas, Nucl. Fusion. 1990. V. 30. P. 1087–1094.
137. *Rej D.J., Barnes G.A., Baron M.H., et al.*, Flux confinement measurements in large field-reversed configuration equilibria, Phys. Fluids. 1990. V. B2. P. 1706–1708.
138. *Hoffman A.L., Slough J.T.*, Field reversed configuration lifetime scaling based on measurements from the large s experiment, Nucl. Fusion. 1993. V. 33. P. 27–38.
139. *Slough J.T., Hoffman A.L., Milroy L.D., Maqueda R., Steinhauer L.C.*, Transport, energy balance, and stability of a large field-reversed configuration. Phys. Plasmas. 1995. V. 2. P. 2286–2291.
140. *Kitano K., Matsumoto H., Yamanaka K., et al.*, Advanced experiments on Field-Reversed Configuration at Osaka, Proc. of 1998 Int. Congress on Plasma Physics & 25th EPS Conf. on Contr. Fusion and Plasma Physics, Prague, 1998.
141. *Steinhauer L.*, FRC data digest, 1996. Data presented in: Iwasawa N., Ishida A., Steinhauer L.C. Tilt mode stability scaling in field-reversed configurations with finite Larmor radius effect, Phys. Plasmas. 2000. V. 7. P.931–934.
142. *Guo H.*, Recent results on Field Reversed Configurations from the Translation, Conformment and Sustainment experiment, Plasma Sci. Technol. 2005. V. 7. P. 2605–2609.
143. *Okada S., Masumoto T., Yamamoto S., et al.*, Recent FRC plasma studies, Fusion Sci. Technol. 2007. V. 51, No. 2T. P. 193–196.
144. *Binderbauer M.W., Tajima T., Steinhauer L.C., et al.*, A high performance field-reversed configuration, Phys. Plasmas. 2015. V. 22. 056110 (16 p.).
145. *Guo H.Y., Binderbauer M.W., Barnes D., et al.*, Formation of a long-lived hot field reversed configuration by dynamically merging two colliding high-β compact toroids, Phys. Plasmas. 2011. V. 18. 056110 (10 p.).
146. *Votroubek G., Slough J., Andreason S., Pihl C.*, Formation of a stable field reversed confguration through merging, J. Fusion Energy. 2008. V. 27. P. 123–127.
147. *Slough J., Votroubek G., Pihl C.*, Creation of a high-temperature plasma through merging and compression of supersonic field reversed configuration plasmoids, Nucl. Fusion. 2011. V. 51. 053008 (10 p.).
148. *Ono Y., Morita A., Katsurai M., Yamada M.* Experimental investigation of three-dimensional magnetic reconnection by use of two colliding spheromaks. Phys. Fluids.

1993. V. B5. P. 3691–3701.

149.   *Ono Y., Matsuyama T., Umeda K., Kawamori E.* Spontaneous and artificial generation of sheared fluw in oblate FRCs in TS-3 and 4 FRC Experiments, Nucl. Fusion. 2003. V. 43. P. 649–654.

150.   *Mozgovoy A.*, Compact toroidal formation in the inductive store, 30th EPS Conference on Contr. Fusion and Plasma Phys., St. Petersburg. ECA. 2003. V. 27A. P. 2.198.

151.   *Slough J.T., Miller K.E.*, Flux generation and sustainment of a field reversed configuration with rotating magnetic field current drive. Phys. Plasmas. 2000. V. 7. P. 1945–1950.

152.   *Hoffman A.L., Guo H.Y., Miller K.E., Milroy R.D.*, Long pulse FRC sustainment with enhanced edge drive magneticfield current drive, Nucl. Fusion. 2005. V. 45. P. 176–183.

153.   *Guo H.Y., et al.*, Improved confinement and current drive of high temperature field reversed configurations in the new translation, confinement, and sustainment upgrade device, Phys. Plasmas. 2008. V. 15. 056101.

154.   *Asai T., Suzuki Y., Yoneda T., et al.*, Experimental evidence of improved confinement in a high-beta field-reversed configuration plasma by neutral beam injection, Phys. Plasmas. 2000. V. 7. P. 2294–2297.

155.   *Okada S., Asai T., Kodera F. et al.*, Experiments on additional heating of FRC plasmas, Nuclear Fusion. 2001. V. 41. P. 625–629.

156.   *Inomoto M., Asai T., Okada S.*, Neutral beam injection heating on field-reversed configurations of plasma decompressrf through axial translation, Nucl. Fusion. 2008. V. 48. 035013 (8 p.).

157.   *Hewlett D.V.*, Spontaneous development of the toroidal magnetic field during the formation of field-reversed theta pinch, Nucl. Fusion. 1984. V. 24. P. 349–357.

158.   *Milroy R.D., Brackbill J.U.*, Toroidal magnetic five generation during compact toroid formation in a field-reversed theta pinch and conical theta pinch, Phys. Fluids. 1986. V. 29. P. 1184–1195.

159.   *Steinhauer L.C.*, Formalism for multiform equilibria with flow, Phys. Plasmas. 1999. V. 6. P. 2734–2741.

160.   *Steinhauer L.C., Guo H. Y.* Nearby-fluids equilibria. II. Zonal flows in a high-β, self-organized plasma experiment, Phys. Plasmas. 2006. V. 13. 052514 (8 p.).

161.   *Wang M.Y., Miley G.H.*, Particle orbits in field-reversed mirrors, Nucl. Fusion. 1979. V. 19. P. 39–49.

162.   *Hsiao M.-Y., Miley G.H.*, Particle-confinement criteria for axisymmetric field-reversed magnetic configurations, Nucl. Fusion. 1984. V. 24. P. 1029–1038.

163.   *Hsiao M.-Y., Miley G.H.*, Velocity-spase particle loss in field-reversed configurations, Phys. Fluids. 1985. V. 28. P. 1440–1449.

164.   *Bozhokin S.V.* About the retention of alpha particles in installations of the compact torus type, Fizika plazmy. 1986. V. 12. P. 1292–1296.

165.   *Khvesyuk V.I., Khvesyuk A.V., Lyakhov A.N.*, Global stochastic particles in a trap with an inverse magnetic configuration, Pis'ma v Zh. Teor. Fiz. 1997. 23. 23. P. 37–39.

166.   *Landsman A.S., Cohen S.A., Glasser A.H.*, Regular and stochastic orbits of ions in a highly prolate field-reversed configuration, Phys. Plasmas. 2004. V. 11. P. 947–957.

167.   *Takahashi T., Inoue K., Iwasawa N., Ishizuka T., Kondoh Y.*, Losses of the neutral beam injected fast ions due to adiabaticity, breaking processes in a field-reversed configuration, Phys. Plasmas. 2004. V. 11. P. 3131–3140.

168.   *Lifschitz A.F., Farengo R., Arista N.R.*, Monte Carlo simulation of a neutral beam injection into a field reversed configuration, Nucl. Fusion. 2002. V. 42. P. 863–875.

169. *Lifschitz A.F., Farengo R., Hoffman A.L.*, Calculations of the tangential neutral beam injection current drive efficiency for present moderate flux FRCs, Nucl. Fusion. 2004. V. 44. P. 1015–1026.

170. *Ferrari H. E., Farengo R.*, Current Drive and Heating by Fusion Protons in a D–³He FRC Reactor, Nucl. Fusion. 2008. V. 48. 035014 (8 p.).

171. *Steinhauer L.C.* Two-dimensional interpreter for field-reversed configurations, Phys. Plasmas. 2014. V. 21. P. 082516.

172. *Steinhauer L.*, Equilibrium rotation in field-reversed configurations, Phys. Plasmas. 2008. V. 15. 012505 (7 p.).

173. *Galkin S.A., Drozdov V.V., Semenov V.N.*, Evolution of the equilibrium of a compact torus with allowance for various loss channels, Plasma Physics. 1989. V. 15. S. 288–299.

174. *Galkin S.A., Drozdov V.V., Semenov V.N.*, A semi-dimensional model of the evolution of equilibrium states of a plasma in a compact torus. Preprint No. 75. IPM named after M.V. Keldysh. Moscow, 1988.

175. *Milroy A. I. D., Barnes D. C., Milroy R. D., Kim C. C., Sovinec C. R.* Simulations of the field-reversed configuration with the NIMROD code, J. Fusion Energy. 2007. V. 26. P. 113–117.

176. *Tuszewski M., Barnes G.A., Baron M.H., et al.*, The n=1 rotational instability in field-reversed configurations, Phys. Fluids. 1990. V. B2. P.2541–2543.

177. *Rej D.J., Taggart D.P., Baron M.H., et al.*, High-power magneticcompression heating of field-reversed configurations, Phys. Fluids. 1992. V. B4. P. 1909–1919.

178. *Ikeyama T., Hiroi M., Ohkuma Y., Nogi Y.*, Detection of electric field around field-reversed configuration plasma, Phys. Plasmas. 2010. V. 17. 012501 (8 p.).

179. *Guo H.Y., Hoffman A.L., Steinhauer L.C., Miller K.E.*, Observations of improved stability and conformation in a high-β self-organized spherical torus-like field-reversed configuration. Phys. Rev. Lett. 2005. V. 95. 175001 (4 p.).

180. *Pustovitov V.D.*, Influence of the current distribution on the stability of the axial region of compact tori, Fizika plazmy. 1981. Vol. 7, No. 5. P. 973–980.

181. *Vabishchevich P.N., Degtyarev L.M., Drozdov V.V., Poshekhonov Yu.Yu., Shafranov V.D.*, About equilibrium configurations in compact tori, Fizika plazmy. 1981. Vol. 7, No. 5. S. 981–985.

182. *Guo H.Y., Hoffman A.L., Milroy R.D., Miller K.E., Votroubek G.R..* Stabilization of interchange modes by rotating magnetic fields. Phys. Rev. Lett. 2005. V. 94. 185001 (4 p.).

183. *Steinhauer L.*, Progress in the physics of reversed magnetic configurations, Izv. VUZ. Mashinostroenie.1999. No. 5–6. P. 106–115.

184. *Nishimura K., Horiuchi R., Sato T.*, Tilt, stabilization by cycling, crossing, magnetic separatrix in a field-reversed configuration, Phys. Plasmas. 1997. V. 4. P. 4035–4042.

185. *Iwasawa N., Ishida A., Steinhauer L. C.* Tilt mode stability scaling in field-reversed configurations with finite Larmor radius effect, Phys. Plasmas. 2000. V. 7. P. 931–934.

186. *Omelchenko Y.A., Schaffer M.J., Parks P.B.*, Nonlinear stability of field-reversed configurations with self-generated toroidal field, Phys. Plasmas. 2001. V. 8. P. 4463–4469.

187. *Belova E.V., Jardin S.C., Ji H., Yamada M., Kulsrud R.*, Numerical study of the global stability of oblate field-reversed configurations, Phys. Plasmas. 2001. V. 8. P. 1267–1277.

188. *Guo H.Y., Hoffman A.L., Miller K.E., et al.*, Achievement of a New High-Confinement, Collisionless FRC State in TCS-Upgrade Improved stability and confinement in a high-β relaxed plasma state, J. Fusion Energy. 2009. V. 28. P. 152–155.

189.   *Krall N.A.*, Low-frequency stability for field reversed configuration parameters, Phys. Fluids. 1987. V. 30, No. 3. P. 878–883.

190.   *Krall N.A.*, The effect of low-frequency turbulence on flux, particle, and energy confinment in a field-reversed configuration, Phys. Fluids. 1989. V. B1. P. 1811–1817.

191.   *Krall N.A.*, Dumping of lower hybrid waves by low-frequency drift waves, Phys. Fluids. 1989. V. B1. P. 2213–2216.

192.   *Sobehart J.R., Farengo R.*, Low-frequency drift dissipative modes in field-reversed configurations, Phys. Fluids. V. B2. 1990. P. 3206–3208.

193.   *Huba J.D., Drake J.F., Gladd T.*, Lower-hybrid-drift instability in field reversed plasmas, Phys. Fluids. 1980. V. 23. P. 552–561.

194.   *Hoffman A.L., Milroy R.D.*, Particle lifetime scalling in field-reversed configurations based on lower-hybrid-drift resistivity. Phys. Fluids. 1983. V. 26. P. 3170–3172.

195.   *Carlson A.W.*, A search for lower-hybrid-drift fluuctuations in a field reversed configuration using CO_2 heterodyne scattering, Phys. Fluids. 1987. V. 30. P. 1497–1509.

196.   *Okada S., Kiso Y., Goto S., Ishimura T.*, Estimation of the electric resistance in field-reversed configuration plasmas from detailed interferometric measurements. Phys. Fluids. 1989. V. B1. P. 2422–2429.

197.   *Okada S., Ueki S., Himura H., et al.*, Measurement of magnetic field in a field-reversed-configuration plasma, Fusion Technol. 1995. V. 27, No. 1T. P. 341–344.

198.   *Farengo R., Guzdar P.N., Lee Y.C.*, Stabilization of the lower hybrid drift modes by finite parallel wavenumber and electron temperature gradients in field-reversed configurations, Phys. Fluids. 1988. V. 31. P. 3299–3304.

199.   *Farengo R., Guzdar P.N., Lee Y.C.*, The effect of magnetized ions on lower hybrid drift instability in field reversed configurations, Phys. Fluids. 1989. V. B1. P. 1654–1658.

200.   *Farengo R., Guzdar P.N., Lee Y.C.*, Collisionless electron temperature gradient-driven instability in field-reversed configurations, Phys. Fluids. 1989. V. B1. P. 2181–2185.

201.   *Okada S., Yamanaka K., Yamamoto S., et al.* Excitation and propagation of low frequency wave in an FRC plasma. Nucl. Fusion. 2003. V. 43. P. 1140–1144.

202.   *Okada S., Inomoto M., Yamamoto S., et al.*, Behavior of a low frequency wave in a FRC plasma, Nucl. Fusion. 2007. V. 47. P. 677–681.

203.   *Iwasawa N., Okada S., Goto S.*, Global eigenmodes of low frequency waves in field-reversed configuration plasmas, Phys. Plasmas. 2004. V. 11. P. 615–624.

204.   *Khvesyuk V.I., Chirkov A.Yu.*, About instabilities in the surface layer of a plasma of an inverted magnetic configuration., Vestnik MGTU. Estestvennye nauki. 2009. No. 1. P. 21–30.

205.   *Khvesyuk V.I., Chirkov A.Yu.*, Peculiarities of Collisionless Drift Instabilities in Poloidal Magnetic Configurations, Plasma Physics Reports. 2010. V. 36, No. 13. P. 1112–1119.

206.   *Chirkov A.Yu., Bendersky L.A., Berdov R.D, Bolshakova AD* Transport model in quasi-equilibrium inverted magnetic configurations, Vestnik MGTU. Estestvennye nauki. 2011. No. 4. P. 15–27.

207.   *Rej D.J., Tuszewski, M.*, A zero-dimensional transport model for field-reversed configurations, Phys. Fluids. 1984. V. 27. P. 1514–1520.

208.   *Steinhauer L.C., Milroy R.D., Slough J.T.*, A model for inferring transport rates from the observed confignment of times in field-reversed configurations, Phys. Fluids. 1985. V. 28. P. 888–897.

209.   *Ikeyama T., Hiroi M., Nogi Y., Ohkuma Y.*, Beta value at separatrix of field-reversed

configuration, Phys. Plasmas. 2009. V. 16. 042512 (5 p.).

210. *Gavrikov M.B., Savel'ev V.V.*, Problems of Plasma Static in Two-Fluid Magnetic Hydrodynamics with Allowance for the Inertia of Electrons, MZhG. 2010. No. 2. P. 176–192.

211. *Suzuki Y., Okada S., Goto S.*, Analysis of averaged β value in two-dimensional equilibria of a field-reversed configuration with end mirror fields, J. Plasma Fusion Res. SERIES. 1999. V. 2. P. 218–221.

212. *Takahashi T., Gota H., Nogi Y.*, Control of elongation for field-reversed configuration plasmas using axial field index of a mirror confinement field, Phys. Plasmas. 2004. V. 11. P. 4462–4467.

213. *Mozgovoy A.G., Romadanov I.V., Ryzhkov S.V.*, Formation of a compact toroid for enhanced efficiency, Physics of Plasmas. 2014. V. 21. 022501.

214. *Romadanov I.V., Ryzhkov S.V., Regimes of pulsed formation of a compact plasma configuration with a high energy input, Plasma Physics Reports. 2015. V. 41. P. 814-819.*

215. *Burtsev V.A., Bozhokin S.V., Dudnikova G.I., et al.*, Quasistationary thermonuclear system based on an inverted magnetic configuration using D–^{3}He-fuel, VANT. Termoyadernyi sintez. 1989. Issue 1. P. 46–52.

216. *Chirkov A.Yu., Khvesyuk V. I.*, Analysis of D–^{3}He / catalyzed D–D plasma as a source of fusion power, Fusion Technol. 2001. V. 39 (1T). P. 406–409.

217. *Kuzenov V.V., Lebo A.I., Lebo I.G., Ryzhkov S.V.*, Physico-mathematical models and methods for calculating the effect of high-power laser and plasma pulses on condensed and gaseous media. Moscow: Bauman Moscow State University. 2015. 328 pp.

218. *Khvesyuk V.I., Chirkov A.Yu.* Parameters of a reactor with an inverted magnetic field in the regime of low-frequency anomalous losses, VANT. Termoyadernyi sintez. 2000. Issue 3. P. 17–27.

219. *Chirkov A.Yu.*, On scaling for the time of confinement of plasmas in an inverted magnetic configuration, Prikladnaya fizika. 2007. No. 2. P. 32–36.

220. *Morozov A.I., Savel'ev V.V.*, Galatean traps with plasma-immersed conductors, Usp. Fiz. Nauk. 1998. P. 168, No. 11. P. 1153–1194.

221. *Kesner J., Mauel M.*, Plasma confinement in a levitated magnetic dipole, Plasma Physics Reports. 1997. V. 23. P. 742–750.

222. *Garnier D.T., Hansen A., Mauel M.E., et al.*, Production and study of high-beta plasma confined by a superconducting dipole magnet, Phys. Plasmas. 2006. V. 13. 056111 (8 p.).

223. *Yoshida Z., Ogawa Y., Morikawa J., et al.*, First Plasma in the RT-1 device, J. Plasma Fusion Res. 2006. V. 1. 008 (2 p.).

224. *Saitoh H., Yoshida Z., Morikawa J., et al.*, Formation of the high-β plasma and stable confinement of toroidal electron plasma in Ring Trap 1, Phys. Plasmas. 2011. V. 18. 056102 (9 p.).

225. *Vovchenko DE, Krashevskaya GV, Kurnaev VA, Khodachenko GV, Zventukh MM* Investigation of plasma pressure profiles in a magnetic trap Magnetor, VANT. Termoyadernyi sintez. 2006. Issue 4. P. 68–76.

226. *Baver D.*, The compact levitated dipole configuration for plasma confinement, J. Fusion Energy. 2011. V. 30. P. 428–432.

227. *Morozov A.I., Bugrova A.I., Bishaev A.M., Kozintseva M.V., Lipatov A.S.*, The concept of myxins for the experimental Galatea reactor, Pis'ma v Zh. Teor. Fiz. 2006. Vol. 32, No. 1. P. 65–70.

228. *Kesner J., Garnier D.T., Hansen A., Mauel M., Bromberg L.*, Helium catalysed D–D fusion in a levitated dipole, Nucl. Fusion. 2004. V. 44. P. 193–203.

229. *Garnier D.T., Kesner J., Mauel M.E.* Magnetohydrodynamic stability in a levitated dipole, Phys. Plasmas. 1999. V. 6. P. 3431–3434.
230. *Kesner J., Simakov A. N., Garnier D. T. et al.* Dipole equilibrium and stability, Nucl. Fusion. 2001. V. 41. P. 301–308.
231. *Furukawa M., Hayashi H., Yoshida Z.*, Stabilization of pressure-driven magneto-hydrodynamic modes by separatrix in dipole plasma confinement, Phys. Plasmas. 2010. V. 17. 022503 (5 p.).
232. *Kesner J.* Stability of electrostatic modes in a levitated dipole, Phys. Plasmas. 1997. V. 4. P. 419–422.
233. *Kesner J.*, Stability of a plasma confined in a dipole field, Phys. Plasmas. 1998. V. 5. P. 3675–3679.
234. *Kesner J., Hastie R. J.*, Electrostatic drift modes in a closed field line configuration, Phys. Plasmas. 2002. V. 9. P. 395–400.
235. *Garnier D.T., Boxer A.C., Ellsworth J.L. et al.* Stabilization of a low-frequency instability in a dipole plasma, J. Plasma Phys. 2008. V. 74. P. 733–740.
236. *Tom R., Tarr J.* Magnetic systems of MHD generators and thermonuclear installations. Moscow: Energoatomizdat, 1985.
237. *Morozov A.I., Nevrovsky V.A., Pistunovich V.I., Svechkopal A.N.* The concept of myxins for the experimental Galatea reactor, Zh. Teor. Fiz. 1999. Vol. 69, No. 4. P. 141–142.
238. *Zhukovsky A., Morgan M., Garnier D., et al.*, Design and fabrication of the cryostat for the filtration of the levitated dipole experiment (LDX), 16th Conf. on Magnet Technology, Tallahassee, Florida, 1999. http://www.psfc.mit.edu/ldx/pubs/zhukovsky_mt16.pdf.
239. *Morozov A.I., Bishaev A.M., Bugrova A.I., Nevrovsky V.A.*, Electric discharge trap Oktupol, Pis'ma v Zh. Teor. Fiz. 1999. Vol. 25, no. 17. P. 57–61.
240. *AI Morozov*. Experimental studies of plasma trap traps in MIREA, VANT. Thermonuclear fusion. 2000. Issue. 3. P. 57–63.
241. *Morozov AI, Chirkov A. Yu.* Analysis of the magnetic configuration of Galatei-3, Materials XXXI Zvenigorod conf. on Plasma Physics and CF, 2004. P. 117.
242. *Lindemuth I. R., Kirkpatrick R. C.* Parameter space for magnetized fuel targets in inertial confinement fusion, Nucl. Fus. 1983. V. 23. P. 263–284.
243. *Hasegawa A., Daido H., Fujita M. et al.* Magnetically insulated inertial fusion: a new approach to controlled thermonuclear fusion, Phys. Rev. Lett. 1986. V. 56. P. 139–142.
244. *Chernychev V. K., Korchagin V. P., Babich L. P. et al.* A Review of Experimental Progress in the MAGO / MTF Thermonuclear Program, IEEE Trans. Plasma Sci. 2016. V. 3. P. 250–267.
245. *Thio Y.C.F., et al.*, Magnetized target fusion in a spherical geometry with stand-off drivers, Proc. II Symposium Current Trends in Int. Fusion Research. 1999. P. 113–134.
246. *Lindemuth I.R., Siemon R.E.*, The fundamental parameter of a controlled thermonuclear fusion space, Am. J. Phys. 2009. V. 77, No. 5. P. 407–416.
247. *Kuzenov V.V., Ryzhkov S.V.*, Developing a procedure for calculating physical processes in combined schemes of plasma magneto-inertial confinement, Bulletin of the Russian Academy of Sciences. Physics. 2016. V. 80, No. 5. P. 598–602.
248. *Kukulin V.I., Voronchev V.T.*, Pinch-based thermonuclear D^3He fusion driven by a femtosecond laser. Phys. Atom. Nucl. 2010. V. 73. P. 1376–1383.
249. *Awe T.J., et al.*, Modified helix-like instability structure on imploding z-pinch liners that are pre-imposed with a uniform axial magnetic field. Phys. Plasmas. 2014. V.

21. 056303.
250. *Feoktistov L.P.*, The future of science. Moscow: Znanie, 1985. P. 168.
251. *Shcherbakov V.A.*, Calculation of the ignition of a thermonuclear laser target by a focusing shock wave, Fizika plazmy. 1983. V. 9. P. 409–411.
252. *Fortov V.E.*, Extreme states of matter on the Earth and in space, Usp. Fiz. Nauk. 2009. V. 179. P. 653–687.
253. *Santarius J. F.*, Compression of a spherically symmetric DT Plasma Liner on a Magnetized DT Target, Phys. Plasmas. 2012. V. 19. 072705.
254. *Garanin S.F./* Physical processes in MAGO-MTF systems. Sarov: F~GUP RFYaTs-VNIIEF, 2012. 342 p.
255. *Azizov E.A., Alikhanov S.G., Galanin M.P., etc.* The project BAIKAL – development of the scheme for generating an electric pulse, VANT. Ser. Termoyadernyi sintez. 2001. No. 3. P. 3–17.
256. *Aleksandrov V.V., Volkov G.S., Grabovsky E.V., et al.* Investigation of the characteristics of implosion of quasispherical wire liners in the Angara-5-1 installation at a current up to 4 MA, Fizika plazmy. 2012. V. 38. P. 345–369.
257. *Chirkov A.Yu., Ryzhkov S.V.*, The plasma jet / laser driven compression of compact plasmoids to fusion conditions. J. Fusion Energy. 2012. V. 31 (1). P. 7.
258. *Kostyukov I.Yu., Ryzhkov S.V.*, Magnetic-inertial thermonuclear fusion with laser compression of a magnetized spherical target, Prikladnaya Fizika. 2011. No. 1. P. 65-72.
259. *Alikhanov S.G., Bakhtin V.P.*, Use of *m*-0 instability of a *z*-pinch liner for three-dimensional plasma compression, DAN SSSR. 1982. V. 263. P. 332–324.
260. *Terki P.J.*, Thermonuclear systems based on θ-pinches with a shrinking liner, Prikl. Mekh. Tekhn. Fiz. 1975. Vol. 4. P. 32–44.
261. *Ivanovskii A.V.*, Realization of magnetic compression in the scheme of the reverse pinch, fed by the current of the VMG," VANT. Termoyadernyi sintez. 2004. Issue 3. P. 37–41.
262. *Azizov E.A., Karev Yu.A., Konkashbaev I.K., et al.* Plasma confinement by an exploding liner, DAN SSSR. 1979. V. 248. P. 1090–1092.
263. *Kurtmullaev R.Kh., Semenov VN, Khvesyuk V.I., et al.*, Plasma accelerators and ion injectors. Moscow: Nauka, 1984. P. 250.
264. *Bogomolov G.D., Velikovich A.L., Liberman M.A.*, On generation of pulsed megagauss fields by compression of a cylindrical liner, Pis'ma v Zh. Teor. Fiz. 1983. Vol. 9, No. 12. P. 748–751.
265. *Artyugina I.M., Zheltov V.A., Komin A.V., et al.* Thermonuclear power plant based on a reactor with a partially evaporating liner, VANT. Ser. Termoyadernyi sintez. 1979. Vol. 1, P. 62–71.
266. *Velikhov E.P., Vedenov A.A., Bogdanets A.D., etc.* On the possibility of creating megagauss magnetic fields by means of a liner compressed by a high-pressure gas, Zh. Teor. Fiz. 1973. V. 43. P. 429–437.
267. *Turchi P. J., et al.*, Rotational stabilization of an imploding liquid cylinder. Phys. Rev. Lett. 1976. V. 36. P. 1546–1549.
268. *Smirnov V.P., Zakharov S.V., Grabovsky E.V.*, Increase of radiation intensity in the quasispherical double liner /dynamic-hohlraum system, Pis'ma v Zh. Eksper. Teor. Fiz. 2005. V. 81, vol. 9. pp. 556–562.
269. *Basko M.M., Kemp A.J., Meyer-ter-Vehn J.*, Ignition conditions for a magnetized target fusion in a cylindrical geometry, Nucl. Fusion. 2000. V. 40. P. 59–68.
270. *Miller R.L.*, Liner-Driven Pulsed Magnetized Target Fusion Power Plant, Fus. Sci. Technol. 2007. V. 52. P. 427–431.

271. *Chang P.Y., Fiksel G., Hohenberger M., et al.* Fusion Yield Enhancement in Magnetized Laser-Driven Implosions, Phys. Rev. Lett. 2011. V. 107. P. 035006.

272. *Rambo P.K., Smith I.C., Porter Jr. J.L. et al.* Z-Beamlet: a multikilojoule, terawatt-class laser system, Applied Optics. 2005. V. 44. P. 2421–2430.

273. *Intrator T.P., Siemon R.E., Sieck P.E.,* Adiabatic model and design of a translating field reversed configuration, Phys. Plasmas. 2008. V. 15. 042505.

274. *Horton R.D., Hwang D.Q., Howard S., et al.* Poloidal field amplification in a coaxial compact toroid accelerator, Nucl. Fusion. 2008. V. 48. 095002.

275. *Eskov A.G., Kozlov N.P., Kurtmullaev R.Kh., et al.* Energy balance in a system with quasispherical liner compression, Pis'ma v Zh. Teor. Fiz. 1983. V. 9. P. 38–41.

276. *Witherspoon F.D., Case A., Messer S.J., et al.,* A contoured gap coaxial plasma with injected plasma armature, Rev. Sci. Instrum. 2009. V. 80. 083506.

277. *Case A., Messer S., Brockington S., et al.,* Merging of high speed argon plasma jets, Phys. Plasmas. 2013. V. 20. 012704.

278. *Uzun-Kaymak I.U., Messer S., Bomgardner R., et al.,* Cross-field plasma injection into mirror geometry, Plasma Phys. Control. Fusion. 2009. V. 51. 095007.

279. *Intrator T.P., Wurden G.A., Sieck P.E., et al.,* Field reversed configuration translation and the magnetized target fusion collaboration, J. Fus. Energy. 2009. V. 28. P. 165–169.

280. *Degnan J.H., Amdahl D.J., Domonkos M., et al.,* Recent magneto-inertial fusion experiments on the field reversed configuration heating experiment, Nucl. Fusion. 2013. V. 53. 093003.

281. *Cassibry J.T., Stanic M., Hsu S.C., et al.,* Ideal hydrodynamic scaling relations for a stagnated imploding spherical plasma liner formed by an array of merging plasma jets, Phys. Plasmas. 2012. V. 19. 052702.

282. *Guo H.Y., Binderbauer M.W., Tajima T.. et al.,* Achieving a long-lived highbeta plasma state by energetic beam injection, Nature Communications. 2015. V. 6. P. 6897.

283. *Laberge M.,* Experimental Results for an Acoustic Driver for MTF, J. Fus. Energy. 2009. V. 28. P. 179–182.

284. *Takeyama S., Kojima E., A* copper-lined magnet coil with a maximum of 700 T for electromagnetic flux compression. J. Phys. D: Appl. Phys. 2011. V. 44. P. 425003.

285. *Gasilov V.A., Dyachenko S.V., Chuvatin A.S., et al.* Analysis of the efficiency of magnetic flux compression by a plasma liner, Matem. modelirovanie. 2009. V. 21. P. 57–73.

286. *Orlov A.P., Repin B.G.,* Numerical simulation of multiwire Z-pinches within the framework of 3-D magnetohydrodynamics, IEEE Trans. Plasma Sci. 2015. V. 8. P. 2515–2519.

287. *Galanin M.P., Lototsky A.P., Rodin A.S., Shcheglov I.A.,* Movement of the liner in the cross section of the magnetic compressor, Vestnik MGTU im. N.E. Baumana. 2010. No. 2. P. 65–84.

288. *Kuzenov V.V., Ryzhkov S.V.,* Numerical modeling of laser target compression in an external magnetic field, Mathematical Models and Computer Simulations. 2018. V. 10, No. 2. P. 255–264.

289. *Ryzhkov S.V.,* The behavior of a magnetized plasma under the action of a laser with high pulse energy, Problems of Atomic Science and Technology. 2010. No. 4 (7). P. 105–110.

290. *Kuzenov V.V., Ryzhkov S.V.* Numerical modeling of magnetized plasma by the laser beams and plasma jets, Problems of Atomic Science and Technology. 2013. No. 1 (83). P. 12–14.

291. *Gotchev O.V., Jang N.W., Knauer J.P., et al.,* Magneto-inertial approach to direct-

drive laser fusion, J. Fus. Energy. 2008. V. 27. P. 25–31.

292.   *Liberman M. ,A., Velikovich A.L.,* Distribution function and diffusion of α-particles in DT fusion plasma, J. Plasma Phys. 1984. V. 31. P. 369–380.

293.   *Felber F.S., Liberman M.A., Velikovich A.L.,* Methods for producing ultrahigh magnetic fields, Appl. Phys. Lett. 1985. V. 46. P. 1042–1044.

294.   *Felber F.S., Malley M.M., Wessel F.J., et al.,* Compression of ultrahigh magnetic fields in a gaspuff Z pinch, Phys. Fluids. 1988. V. 31. P. 2053–2056.

295.   *Gotchev O.V., Chang P.Y., Knauer J.P., et al.,* Laser-driven magnetic flux compression in high-energy-density plasmas, Phys. Rev. Lett. 2009. V. 103. 215004.

296.   *Perkins L.J., Logan B.G., Zimmerman G.B., et al.,* Two-dimensional simulations of the thermonuclear fusion targets in the case of compressed axial magnetic fields, Phys. Plasmas. 2013. V. 18. 072708.

297.   *Knauer J.P., Gotchev O.V., Chang P.Y., et al.,* Compressing magnetic fields with high-energy lasers, Phys. Plasmas. 2010. V. 17. 056318.

298.   *Ryzhkov S.V.,* Current status, problems and prospects of thermonuclear facilities based on the magneto-inertial confinement of hot plasma. Bulletin of the Russian Academy of Sciences. Physics. 2014. V. 78, No. 5. P. 456–461.

299.   *Ivanovskii A.V.,* Realization of magnetic compression in the scheme of the reverse pinch, fed by the current of VMG," in: Problemy atomnoi nauki i tekhniki. Termoyadernyi sintez. 2004. Issue 3. P. 37–41.

300.   *Aleksandrov V.V., Gasilov V.A., Grabovsky E.V., et al.,* Increase in the energy density in the pinch plasma under three-dimensional compression of quasi-spherical wire liners, Fizika plazmy. 2014. V. 40, No. 12. P. 1057–1073.

301.   *Kuzenov V.V., Ryzhkov S.V.,* Radiation-hydrodynamic modeling of the contact boundary of the plasma target placed in an external magnetic field, Applied Physics. 2014. No. 3. P. 26–30.

302.   *Rahman H.U., Wessel F. J., Ney P., et al.,* Shock waves in a Z-pinch and the formation of a high energy density plasma, Phys. Plasmas. 2012. V. 19. 122701.

303.   *Cassibry J.T., Stanic M., Hsu S.C.,* Ideal hydrodynamic scaling relations for a stagnated imploding spherical plasma, Phys. Plasmas. 2013. V. 20. 032706.

304.   *Hsu S.C., Moser A.L., Merritt E.C., et al.,* Laboratory plasma physics experiments using merging supersonic plasma jets, Journal of Plasma Physics. 2015. V. 81. 345810201.

305.   *Slutz S.A., Stygar W.A., Gomez M.R., et al.,* Scaling magnetized liner inertial fusion on Z and future pulsed-power accelerators, Phys. Plasmas. 2016. V. 23. 022702.

306.   *Ryzhkov S.V.,* A field-reversed magnetic configuration and applications of high-temperature FRC plasma, Plasma Physics Reports. 2011. V. 37, No. 13. P. 1075–1081.

307.   *Kuzenov V.V., Lebo A.I., Lebo I.G., Ryzhkov S.V.,* Physico-mathematical models and methods for calculating the effect of high-power laser and plasma pulses on condensed and gaseous media (2nd ed.). Moscow: BMSTU. 2017. 328 p.

308.   *Matsuda Y.H., Herlach F., Ikeda S., et al.,* Generation of 600 T by electromagnetic flux compression with improved implosion symmetry, Rev. Sci. Instrum. 2002. V. 73. P. 4288–4294.

309.   *Votroubek G., Slough J.,* The Plasma Liner Compression Experiment, J. Fus. Energy. V. V. P. P. 571.

310.   *McBride R.D., et al.,* Penetrating Radiography of Imploding and Stagnating Beryllium Liners on the Z Accelerator, Phys. Rev. Lett. 2012. V. 109. 135004.

311.   *Siemon R.E.* Atlas for magneto-inertial fusion, 6th Symp. Current Trends Int. Fusion Res.: a review. 2005. URL: http:,www.physicsessays. com / doc / s2005 / Richard_E_Siemon_talk_mar_2005.pdf (circulation date 09.10.2011).

312.   *Ryzhkov S.V., Chirkov A.Yu., Ivanov A.A.,* Analysis of the compression and heating

of magnetized plasma targets for magneto-inertial fusion, Fus. Sci. Technol. 2013. V. 63 (1T). P. 135–138.

313. *Samulyak R., Parks P., Wu L.,* Spherically symmetric simulation of plasma liner driven magneto-inertial fusion, Phys. Plasmas. 2010. V. 17. 092702.

314. *Dmitriev V.F.,* Influence of polarization on the cross section and angular distributions of the reaction products ^{11}B (p, α) ^{8}Be*, Yadernaya fizika. 2006. V. 69. P. 1496–1497.

315. *Cuneo M.E., et al.,* Magnetically driven implosions for inertial confinement fusion at Sandia National Laboratories, IEEE Trans. Plasma Sci. 2012. V. 40. P. 3222–3245.

316. *McBride, et al.* Beryllium liner implosion experiments on the Z accelerator in preparation for magnetized liner inertial fusion, Phys. Plasmas. 2013. V. 20. 056309.

317. *Laberge M.,* An acoustically driven magnetized target fusion reactor. J. Fusion Energy. 2008. V. 27. P. 65–68.

318. *Nakamura D., et al.,* Experimental evidence of threedimensional dynamics of an electromagnetically imploded liner, Rev. Sci. Instrum. 2014. V. 85. P. 036102.

319. *Gasilov V.A., et al.,* Numerical modeling of compression of toroidal plasma by a quasispherical liner. M., 1979. 36 p. (IPM M.V. Keldysha Preprint No. 71).

320. *Orlov A.P., Repin B.G.,* Three-dimensional magnetohydrodynamic modeling of the implosion of multiwire cylindrical liners using the FLUX-3D program., Matem. modelirovanie. 2014. Vol. 26. P. 3–16.

321. *Gasilov V.A., et al.* A package of MARPLE3D applications for modeling high-performance computers of pulsed magnetically accelerated plasma, Matem. modelirovanie. 2012. V. 24. P. 55–87.

322. *Parks P.B.,* A model of cusp magnetic field compression by an expanding plasma fireball, Phys. Plasmas. 2005. V. 12. 102510 (7 p.).

323. *Khariton Yu.B., et al.,* On the work of thermonuclear targets with magnetic compression, Usp. Fiz. Nauk. 1976. V. 120. 706.

324. *Murakami M., Nishihara K.,* Efficient shell implosion and target design, Japanese Journal of Applied Physics. 1987. V. 26. P. 1132–45.

325. *Robson A.E.,* The flying cusp: a compact pulsed fusion system, Naval Research Laboratory Report MR-2692. 1973. 26 p.

326. *Pfalzner S.,* An Introduction to Inertial Confinement Fusion, Series in Plasma Physics. Taylor & Francis: New York-London, 2006. 232 p.

327. *Spalding I.J.,* Cusp confinement and thermonuclear reactors, Nuclear Fusion. 1968. V. 8. P. 161–171.

328. *Kidder R.E.,* Laser-driven compression of hollow shells: power requirements and stability limitations. Nuclear Fusion. 1976. V. 16. P. 3–14.

329. *Kidder R. E.,* Energy gain of laser-compressed pellets: a simple model calculation. Nuclear Fusion. 1976. V. 16. P. 405–408.

330. *Atzeni S., Meyer-ter-Vehn J.* The physics of inertial fusion: beam plasma interaction, hydrodynamics, hot dense matter. Oxford University Press: Oxford, 2009. 356 p.

331. *Karbushev D.N., Ryzhkov S.V., and Troyanik M.K.,* On Improved Analytical Dependences for Energy Release Rates and Synthetic Reactions Cross sections, Nauka i obrazovanie: Electronic Scientific and Technical Publication. 2009. T. 4. URL: http://technomag.edu.ru/doc/117768.html (circulation date on October 9, 2011)

332. International Atomic Energy Agency. Department of Nuclear Sciences & Applications. Division of Physical & Chemical Sciences. Fusion charged particle sublibrary: FENDL / C-2.0. URL: http://www-nds.iaea.org/fendl/fenfusion.htm (reference date 09.10.2011)

333. *Guskov S.Yu., et al.,* Energy transfer by charged particles in a laser plasma, Kvant. elektronika. 1974.No. 7. P. 1617–1623.

334. *Chirkov A. Yu., et al.,* Fusion modes of an axially symmetric mirror of a trap with the high power injection of fast particles, Plasma Phys. Rep. 2012. V. 38. P. 1025–1031.
335. *Solov'ev L.S.,* Hydromagnetic stability of closed plasma configurations, in: Problemy teorii plazmy. Moscow: Atomizdat, 1972. Vol. 6. P. 210-290.
336. *Lamb G.* Hydrodynamics, Leningrad: OGIZ, 1947. 929 p.
337. *Steinhauer L.C.,* Improved analytic equilibrium for a field-reversed configuration, Phys. Fluids. 1990. V. B2, No. 12. P. 3081–3085.
338. *Bruno S.,* Magneto-inertial fusion and magnetized HED physics, Workshop on Scientific Opportunities in High Energy Density Plasma Physics. Washington DC, 2008. URL: http://fsc.lle.rochester.edu/hedlp/panelmeetings. php (circulation date 09.10.2011).
339. *Kuzenov V.V., Ryzhkov S.V.,* Evaluation of hydrodynamic instabilities in inertial confinement fusion target in a magnetic field, Problems of Atomic Science and Technology. 2013. No. 4 (86). P. 103–107.
340. *Ponchaut N.F., et al.,* On imploding cylindrical and spherical shock waves in a perfect gas, Journal of Fluid Mechanics. 2006. V. 560. P. 103–122.
341. *Ryutov D.D., et al.,* Particle and heat transport in a dense wall-confined MTF plasma (theory and simulations), Nuclear Fusion. 2003. V. 43. P. 955–960.
342. *Intrator T.,* MTF users perspective: what is needed from a compression scheme, Plasma Jet Workshop. LANL, 2008. URL: http://wsx.lanl.gov/ Plasma-Jet-Workshop-08 / intrator.pdf (circulation date 09.10.2011)
343. *Frank-Kamenetskii D.A.,* Lectures on plasma physics. Moscow: Atomizdat, 1968. 287 p.
344. *Braginskii, SI,* Phenomena of transport in a plasma, in: Problemy teorii plazmy. Moscow: Gosatomizdat, 1963. Vol. 1. P. 183–272.
345. ECPL-82. The LLNL Evaluated Charged-Particles Data Library, Summary of Contents in: Documentation Series IAEA-NDS-56. 1983. IAEA Nuclear Data Section, Vienna, Austria.
346. *Feldbacher R.,* Nuclear reaction cross sections and reactivity parameter library and files. Vienna: IAEA, 1987.
347. *Nevins W.M., Swain R.,* The thermonuclear fusion rate coefficient for p–^{11}B reactions, Nuclear Fusion. 2000. V. 40. P. 865–872.
348. *Kastler A.,* J. Phys. Radium. V. 11. 1950. P. 255–265.
349. *Happer W.,* Optical pumping, Rev. Mod. Phys. V. 44. 1972. P. 169–249.
350. *Suter D.,* The physics of laser-atom interactions. Cambridge University Press, 1997. 457 p.
351. *Coulter K.P., et al.* Spin-exchange optical pumping as a source of spin-polarized atomic deuterium, Phys. Rev. Lett. V. 68. 1992. P. 174-177.
352. *Stenger J., et al.,* First experimental verification of spin temperature in high flow spin-exchange source for polarized H atoms, Phys. Rev. Let. 1997. V. 78. P. 4177–4180.
353. *Paetz Schieck H.,* Experiments on four-nucleon reactions, Few Body Syst. 1988. V. 5. P. 171.
354. *Leonard D.S., et al.,* Precision measurements of H2 (d, p) H3 and H2 (d, n) He3 total cross sections at Big Bang nucleosynthesis energies, Phys. Rev. C. Nuclear Physics. 2006. V. 73. P. 045801.
355. *Glockle W., et al.,* The three-nucleon continuum: Achievements, challenges and applications. Phys. Rep. 1996. V. 274. P. 107–285.
356. *Gojuki S., Oryu S.* Polarization effects in the ^{3}He (d, p) ^{4}He fusion reaction, Modern Physics Letters A. 2003. V. 18. P. 302–305.

Index

Symbols

θ-pinch 63, 76, 107

A

Apollo 26, 27
ARIES-III 26, 27
ARTEMIS project 93, 166

B

Barnes beta 64
boron-11 4
boundary
  Ohkawa boundary 98, 99, 100, 104
bremsstrahlung 7, 8, 10, 20, 23, 26, 38, 39, 57, 91, 128, 137, 142, 143

C

coefficient
  coefficient of dynamic friction 55
  Troyon coefficient 37
Compact Toroid Challenge 76
compression
  isoentropic compression 120
configuration
  compact configuration 5, 77, 80, 95
  compact toroidal configuration 62
  dipole configuration 95, 96, 101, 103
  field reversed configuration 5, 61
  magnetic configuration 5, 23, 26, 28, 29, 31, 32, 47, 65, 69, 73, 78, 85,
      86, 93, 94, 98, 99, 100, 102, 109, 110, 115, 121, 126, 131
  multipole configuration 95, 101, 102, 103, 104
  Trimix' configuration 101
confinement

Milton Keynes UK
Ingram Content Group UK Ltd.
UKHW040057071024
449327UK00019B/628